Christoph Kurfürst

Cryogenic Beam Loss Monitoring for the LHC

Christoph Kurfürst

Cryogenic Beam Loss Monitoring for the LHC

Investigations on a possible upgrade for the LHC machine protection

Südwestdeutscher Verlag für Hochschulschriften

Impressum / Imprint

Bibliografische Information der Deutschen Nationalbibliothek: Die Deutsche Nationalbibliothek verzeichnet diese Publikation in der Deutschen Nationalbibliografie; detaillierte bibliografische Daten sind im Internet über http://dnb.d-nb.de abrufbar.
Alle in diesem Buch genannten Marken und Produktnamen unterliegen warenzeichen-, marken- oder patentrechtlichem Schutz bzw. sind Warenzeichen oder eingetragene Warenzeichen der jeweiligen Inhaber. Die Wiedergabe von Marken, Produktnamen, Gebrauchsnamen, Handelsnamen, Warenbezeichnungen u.s.w. in diesem Werk berechtigt auch ohne besondere Kennzeichnung nicht zu der Annahme, dass solche Namen im Sinne der Warenzeichen- und Markenschutzgesetzgebung als frei zu betrachten wären und daher von jedermann benutzt werden dürften.

Bibliographic information published by the Deutsche Nationalbibliothek: The Deutsche Nationalbibliothek lists this publication in the Deutsche Nationalbibliografie; detailed bibliographic data are available in the Internet at http://dnb.d-nb.de.
Any brand names and product names mentioned in this book are subject to trademark, brand or patent protection and are trademarks or registered trademarks of their respective holders. The use of brand names, product names, common names, trade names, product descriptions etc. even without a particular marking in this work is in no way to be construed to mean that such names may be regarded as unrestricted in respect of trademark and brand protection legislation and could thus be used by anyone.

Coverbild / Cover image: www.ingimage.com

Verlag / Publisher:
Südwestdeutscher Verlag für Hochschulschriften
ist ein Imprint der / is a trademark of
OmniScriptum GmbH & Co. KG
Heinrich-Böcking-Str. 6-8, 66121 Saarbrücken, Deutschland / Germany
Email: info@svh-verlag.de

Herstellung: siehe letzte Seite /
Printed at: see last page
ISBN: 978-3-8381-3919-7

Zugl. / Approved by: Wien, TU, Dissertation, 2013

Copyright © 2015 OmniScriptum GmbH & Co. KG
Alle Rechte vorbehalten. / All rights reserved. Saarbrücken 2015

Abstract

A Beam Loss Monitoring (BLM) system was installed on the outside surface of the LHC magnet cryostats to protect the accelerator equipment from beam losses. The protection is achieved by extracting the beam from the ring in case thresholds imposed on measured radiation levels are exceeded.

Close to the interaction regions of the LHC, the present BLM system is sensitive to particle showers generated in the interaction region of the two beams. In the future, with beams of higher energy and brightness resulting in higher luminosity, distinguishing between these interaction products and possible quench-provoking beam losses from the primary proton beams will be challenging. The particle showers measured by the present BLM configuration are partly shielded by the cryostat and the iron yoke of the magnets. The system can hence be optimised by locating beam loss monitors as close as possible to the protected element, i. e. the superconducting coils, inside the cold mass of the magnets in superfluid helium at 1.9 K. The advantage is that the dose measured by the Cryogenic Beam Loss Monitor (CryoBLM) would more precisely correspond to the dose deposited in the superconducting coil.

The main challenges of this placement are the low temperature of 1.9 K and the integrated dose of 2 MGy in 20 years. Furthermore the CryoBLM should work in a magnetic field of 2 T and at a pressure of 1.1 bar, withstanding a fast pressure rise up to 20 bar in case of a magnet quench. The detector response should be linear between 0.1 and 10 mGy/s and faster than 1 ms. Once the detectors are installed in the LHC magnets, no access will be possible. Hence the detectors need to be available, reliable and stable for 20 years. Following intense research it became clear that no existing technology was proven to work in such conditions. The candidates under investigation in this work are diamond and silicon detectors and an ionisation chamber, using the liquid helium itself as particle detection medium.

All the selected detector technologies are based on ionisation and subsequent charge carrier transport within the detector bulk. Therefore laboratory measurements were performed to measure the charge carrier characteristics in the detector material in the temperature range from 1.6 to 300 K. In the silicon detector, charges were generated using laser light and α-particles. For diamond detectors the measurements were done with α-particles only. The temperature dependence of the drift velocity and of the mobility of the charge carriers was measured.

To measure the detector's characteristics with respect to particle detection at liquid helium temperatures, low intensity beam tests with minimum ionising protons were carried out. They allowed to prove that all tested detectors work at 1.9 K. The silicon detector Full Width Half Maximum (FWHM) of the signal from a MIP is 2.5 $\pm$ 0.7 ns at liquid helium temperatures. For the diamond detector the FWHM is 3.6 $\pm$ 0.8 ns. The signal width decrease from room temperature to liquid helium temperatures is of 54 % for silicon material and 28 % for diamond material. This allows bunch by bunch resolution of the LHC losses, as already demonstrated at room temperature.

The radiation hardness of the solid-state detectors over 20 years of LHC operation was addressed during high intensity beam tests carried out at CERN in a liquid helium environment. A complete cryogenic system was installed in the irradiation area of the CERN East Hall. Data from the continuous monitoring of the signal development during irradiation and measurements from test cycles enabled the advantages and disadvantages of each detector technology to be identified. The expected reduction in detector sensitivity over 20 years (2 MGy) of LHC operation is of a factor of 14 $\pm$ 3 for the diamond detector. For the silicon detector the expected signal reduction is of a factor of 25 $\pm$ 5.

Using liquid helium as particle detection medium has the advantage of no radiation hardness issues. The downside is the low electron and ion mobility in superfluid helium, which leads to a slower detector response. With the current design of the liquid helium chamber a successful protection from losses with a time constant above 180 µs is ensured.

These results show that the diamond and silicon detectors satisfy the criteria for use in a fast protection and feedback system, while the simultaneous use of the liquid helium chamber enables the calibration of the solid-state detectors and the reliable protection from steady state losses.

Kurzfassung

Der am CERN aufgebaute Large Hadron Collider (LHC) ist der weltweit leistungsstärkste Teilchenbeschleuniger. Strahlverlustmonitore (BLMs) wurden außerhalb der LHC Magnetkryostaten entlang des Ringes installiert, um die Elemente des Beschleunigers vor Strahlenverlusten zu schützen. Direkte Verluste des Strahles können unter anderem den unerwünschten Übergang der supraleitenden Magnete zum normalleitenden Zustand zur Folge haben. Der Schutz wird gewährleistet, indem der Strahl extrahiert wird im Falle zu hoher Teilchenverluste.

Die derzeitigen Strahlverlustmonitore des LHC sind im Bereich der Interaktionspunkte des LHC empfindlich auf Teilchenschauer der Kollisionen zwischen den Strahlen. Die Differenzierung zwischen diesen Kollisionsprodukten und direkten Verlusten des primären Strahles wird in Zukunft, für den Hochraten-LHC, mit höherer Strahlenergie und -Intensität, herausfordernd, da die Teilchenschauer zum Teil durch den Kryostaten und das Jocheisen der Magnete abgeschirmt sind. Der Schutz kann gewährleistet bleiben, indem die Strahlverlustmonitore dichter zum Strahl und zum Magneten installiert werden. Das entspricht einer Installation der Detektoren im suprafluiden Helium bei 1.9 K. Der dadurch gewonnene Vorteil ist, dass das gemessene Signal der kryogenen Detektoren (CryoBLM) mit der Energiedeposition im Magneten besser übereinstimmt.

Die Hauptherausforderungen dieser Platzierung sind die Temperatur von 1.9 K und die integrierte Strahldosis von 2 MGy in 20 Jahren. Desweiteren muss der kryogene Detektor in einem Magnetfeld von 2 T und einem Druck von 1.1 bar funktionieren. Er muss einer möglichen Drucksteigerung auf 20 bar standhalten, sollte der Übergang eines Magneten zum normalleitenden Zustand stattfinden. Nach der Installation der kryogenen Detektoren in den Magneten des LHC, sind diese nicht mehr erreichbar und müssen daher über 20 Jahre stabil und betriebssicher sein.

Nach intensiven Recherchen, wurde klar, dass bisher für keine Detektortechnologie ein Funktionieren unter diesen Konditionen bewiesen wurde. Die in dieser Arbeit untersuchten Kandidaten sind Diamant- und Siliziumdetektoren und eine auf flüssigem Helium basierende Ionisationskammer.

Um den Ladungstransport im Detektormaterial im Kalten zu ermitteln, wurden erste Messungen im Labor durchgeführt. Ladungen konnten im Siliziumdetektor mittels Laserlicht und Alphateilchen erzeugt werden. Für Diamantdetektoren wurden die Labormessungen nur mittels Alphateilchen durchgeführt. Die Temperaturabhängigkeit der Driftgeschwindigkeit und der Mobilität der Ladungen konnte gemessen werden.

Um die Detektoreigenschaften für Einzelteilchenmessungen bei Temperaturen im flüssigen Helium zu ermitteln, wurden Strahltests mit minimal ionisierenden Protonen durchgeführt. Die Signalbreite nimmt von Raumtemperatur zu Temperaturen im flüssigen Helium um 54 % im Siliziummaterial ab und um 28 % im Diamantmaterial. Dies erlaubt eine "bunch by bunch" Auflösung der LHC-Strahlverluste, wie bereits bei Raumtemperatur erfolgreich gezeigt werden konnte.

Die Strahlenhärte der Festkörperdetektoren über 20 Jahre LHC Betrieb wurde im Zuge von Bestrahlungen in flüssigem Helium mit einem Protonenstrahl hoher Intensität gemessen. Ein vollständiges kryogenes System wurde in der Bestrahlungszone installiert. Die Daten der Entwicklung der Ladungsausbeute während der Bestrahlung und die Messungen von Testzyklen erlauben es die Vor- und Nachteile der Detektortechnologien zu identifizieren. Die Reduktion der Detektorempfindlichkeit über 20 Jahre (2 MGy) LHC Betrieb beträgt einen Faktor 14 ± 3 für den Diamantdetektor, während für den Siliziumsensor der Faktor 25 ± 5 ist.

Die Verwendung von flüssigem Helium als Teilchendetektionsmedium hat den Vorteil, dass Strahlenhärte kein Problem darstellt. Der Nachteil ist die langsame Ladungsdrift, aufgrund der speziellen Struktur der Ladungen im flüssigen Helium. Mit der derzeitigen Konzeption der Ionisationskammer kann ein erfolgreicher Schutz vor Strahlverlusten mit einer Zeitkonstante größer als 180 µs gewährleistet werden.

Die gewonnen Erkenntnisse motivieren den Einsatz der Diamant- und Siliziumdetektoren als schnelles Schutz- und Feedbacksystem, während das zeitgleiche Messen mittels Heliumionisationskammer zur Kalibrierung der Festkörperdetektoren und zum Schutz vor stationären Strahlverlusten verwendet werden kann.

Acknowledgements

During this project I have been challenged many times on many different levels. Without the people I mention here, it would have never been possible for me to organise and perform all these experiments, to keep the measurement conditions stable at all time, to analyse the data and to obtain and interpret the results, that I am able to present in this work.

I want to thank **Prof. Christian Fabjan**, who agreed to supervise me and the project. He was the one most interested in the progress of the liquid helium chamber. While others were focussed on my results with the silicon and diamond detectors, he motivated me to find out more about the liquid helium particle detection properties. Through his experience with liquid detectors and past work with the BLM section, he could optimally guide and follow my progress. I am grateful for his support.

I want to thank **Mariusz Sapinski**, who strengthened my back in the moments I was overwhelmed by the enormity of the tasks. He was of great support to me and helped me many times in many ways, even for little scientific tasks like pulling cables or refilling cryostats . He is the father of the project, motivating the investigation of CryoBLMs, based on the simulation results of **Alessio Mereghetti**. **Bernd Dehning** was of great help and support. He was very interested in the project and in my results. He freed my way to get a maximum of necessary equipment and instruments, enabling the best possible measurements. The discussions with him depicted for me weak argumentation, essential points of the project and enhanced my progression. Together with **Rohdri Jones** they further enabled me to participate in several workshops and conferences, where I could discuss the project with specialists. **Barbara Holzer** agreed on selecting me as one of her students for the BLM section. Her questions and remarks during meetings were always helpful. The meetings she organises allowed me to improve my presentation skills. I especially remember her giving me useful tips when I got selected for a presentation at a conference with 1320 participants.

Prof. Erich Griesmayer invested many extra hours and material (diamonds and amplifiers) with his company CIVIDEC into the project and hence also into my progress. I highly appreciated and followed his advice in many different areas of my private and professional life. **Christina Weiss** helped me in the execution of two different beam tests. I could profit from our scientific discussions and her advanced amplifier knowledge. Further acknowledgements go to **Sophie Mallows** and **Jacobus Van Hoorne**, who shared the beam with me two times during which we could help each other.

Vladimir Eremin and **Elena Verbitskaya**, from the Ioffe Institute in St. Petersburg, were of great help. I highly appreciate having had the opportunity to visit their institute and learn from them. Together with **Jaakko Haerkoenen**, the spokesperson of the **RD39** collaboration, they provided me with silicon detectors for testing, silicon knowledge as well as opportunities to learn, discuss and present my results.

Thomas Eisel played a central role in the cryogenic challenges for the CryoBLM team. All major cryogenic preparations, questions and issues went through his hands, in the laboratory, during beam tests and during irradiation. **Carlos Arregui Rementeria** helped Thomas and me a lot, especially in the preparation and execution of the cryogenic irradiation. I further would like to point out **Gerhard Burghardt**, who was so nice to help me during weekends, even in the middle of the night, with cryogenic manipulations, although he had a baby at home of less than one month of age. Further acknowledgements to all the members of the cryogenic laboratory that helped at some point with the CryoBLM project.

I would like to thank **Marcin Bartosik** for his help and discussions during the cryogenic irradiation tests and the LHC detector installations.

Lau Gatignon, as beam line responsible, gave me the opportunity to perform the beam tests and was very helpful in the introduction to all information and instrumentation concerning the beam and the test area. **Maurice Glaser** is responsible for the irradiation area. His 20 years of experience in the area, his knowledge and his pragmatism played a major role in the success of the preparation and performance of the irradiation tests. His colleague **Federico Ravotti** was also of great help and still is of great help in dealing with all organisational issues with past and future irradiation tests as well as with irradiated material.

I thank **Raymond Tissier**, who I want to mention especially with respect to the outstanding solutions he found within little time for the construction of the irradiation tables. I also want to acknowledge the work done by our colleague **Ion Savu**, who was involved in the construction of the many different detector supports for the experiments.

Morad Hamani and Frederic Camba helped in the careful construction of three liquid helium chamber prototypes, optimised for high voltage and cryogenic temperatures.

I thank **Michael Moll** for taking the time to discuss the radiation hardness of silicon detectors and **Werner Riegler** for discussions on the liquid helium chamber properties.

Elisa Guillermain is an optical fibre specialist and supported the project with her input with respect to radiation hard fibres and the complications related to cryogenic installations.

I want to thank **Heinz Pernegger** and **Hendrik Jansen** as diamond material specialists for discussions about diamond detectors and their particular properties. **Moritz Guthoff** and the **RD42** collaboration were of great help for me to discuss and understand the radiation hardness characteristics of diamond detectors.

I thank **Robert Froeschl** for radiation protection relevant simulations and the preparation of the resulting values in the safety reports. **Bruno Pichler** is a safety expert and prepared me with respect to my responsibility in safety relevant issues during the execution of beam tests for all participating personnel.

Ewald Effinger was of great help for questions concerning the optimisation of the design of the readout system of the different detectors.

I thank **Christian Boccard** for his advice and for his material that we needed for the installation of the first cryogenic LHC BLMs. I want to thank **Vittorio Parma**, **Thierry Renaglia** and **Michel Souchet**, who helped and enabled the installation of detectors on the LHC magnets. It meant a lot of extra work for them and especially for LHC magnet people, where the pressure is high and the schedule is tight.

Further acknowledgements to all the members of the BLM section and the BI group that supported the CryoBLM efforts continuously. I appreciated spending time with them during coffee breaks, meetings and seminars.

Finally major thanks go to my family. They did not appreciate my absence, but I received anyhow a lot of support and encouragement from them, throughout my entire career.

Thank you. I wish you all the best for your future.

Abbreviations

AC:	Alternating Current
ALICE:	A Large Ion Collider Experiment
ATLAS:	A Toroidal LHC ApparatuS
BLM:	Beam Loss Monitor
BPM:	Beam Position Monitor
CCD:	Charge Collection Distance
CCE:	Charge Collection Efficiency
CERN:	Centre Européen pour la Recherche Nucléaire
	(European Organisation for Nuclear Research)
CID:	Current Injected Detector
CMS:	Compact Muon Solenoid
CryoBLM:	Cryogenic Beam Loss Monitor
CSA:	Charge Sensitive Amplifier
CVD:	Chemical Vapour Deposition
DC:	Direct Current
EUV:	Extreme Ultra Violet
FWHM:	Full Width Half Maximum
IP:	Interaction Point
I-V:	Current Voltage
LHC:	Large Hadron Collider
LHCb:	Large Hadron Collider beauty
LHe:	Liquid Helium
MFP:	Mean Free Path
MIP:	Minimum Ionising Particle
NIEL:	Non-Ionising Energy Loss
NPLC:	Number of Power Line Cycles
pCVD:	polycrystal Chemical Vapour Deposition diamond detector
PS:	Proton Synchrotron
QPS:	Quench Protection System
RMS:	Root Mean Square
RT:	Room Temperature
SCR:	Space Charge Region
sCVD:	Single crystal Chemical Vapour Deposition diamond detector
SEC:	Secondary Emission Chamber
Si:	Silicon detector
TCT:	Transient Current Technique
TES:	Transition Edge Sensor
UFO:	Unidentified Falling Object

Contents

1 **Introduction**   11

2 **The LHC and its BLM system**   13
   2.1 LHC (Large Hadron Collider) 13
   2.2 The experiments and their goals 14
   2.3 Risks from beam for LHC equipment 16
     2.3.1 Beam loss origins 16
     2.3.2 Impact of beam losses on accelerator elements 17
     2.3.3 Machine protection 19
   2.4 The LHC Beam Loss Monitoring (BLM) system 20
     2.4.1 The BLM Ionisation Chambers 20
     2.4.2 Quench preventing thresholds 21
     2.4.3 The limits of the present BLM system 24

3 **The cryogenic BLM project**   27
   3.1 CryoBLM as solution for the present BLM limitation 27
     3.1.1 Requirements for the cryogenic BLM application 27
   3.2 Considered detector technologies 29
     3.2.1 Gaseous Helium3 ionisation chamber 29
     3.2.2 Thermal equilibrium calorimeters 29
     3.2.3 Liquid helium scintillation light 30
     3.2.4 Transition edge sensors 32
     3.2.5 Scintillators 33
   3.3 Selected detector technologies 33
     3.3.1 Solid-state particle detectors 33
     3.3.2 Liquid helium ionisation chamber 36
   3.4 Summary 38

4 **Charge carrier characteristics at cryogenic temperatures**   40
   4.1 The transient current technique 40
   4.2 Cryogenic basics 42
   4.3 Charge carrier characteristics in silicon material 42
     4.3.1 Experimental setup 42

	4.3.2	Reverse bias measurements	44
	4.3.3	Silicon detector forward bias operation	50
	4.3.4	Infra-red light pulses	52
4.4		Charge carrier characteristics in diamond material	54
4.5		Charge carrier characteristics in liquid helium	56
	4.5.1	Positive Ions	56
	4.5.2	Negative Ions	57
	4.5.3	Electron-ion recombination	58
	4.5.4	Breakdown field in liquid helium	58
	4.5.5	Estimation of the ionised charge per MIP	58
	4.5.6	Estimation of the charge collection time	59
	4.5.7	Magnetic field considerations	60
4.6		Summary	61

5 Cryogenic single particle detection with diamond and silicon detectors — 62

5.1		Signal estimation from a MIP	62
5.2		Fluctuations of energy loss - Landau distribution	63
	5.2.1	Effects of cuts on the Landau distribution	63
5.3		Cryogenic amplifier	64
	5.3.1	Introduction	64
	5.3.2	Testing procedure	65
	5.3.3	Electronic noise variation	66
	5.3.4	Gain variation	66
	5.3.5	Power consumption	68
	5.3.6	Further measurements	68
	5.3.7	Cryogenic amplifier summary	70
5.4		Single particle detection	70
	5.4.1	Experimental setup	70
	5.4.2	Beam line	70
	5.4.3	Results	73
5.5		Summary	78

6 Radiation hardness of diamond and silicon detectors — 80

6.1		Radiation hardness	80
	6.1.1	Derivation of the charge collection degradation model	81
6.2		Irradiation facility and beam properties	83
6.3		Measurements for radiation hardness characterisation	83
	6.3.1	Option 1: Energy loss measurements	84
	6.3.2	Option 2: Silicon detector TCT using a pulsed laser	85
	6.3.3	Option 3: Diamond detector TCT using an alpha-particle source	85
	6.3.4	Option 4: DC measurements	86
	6.3.5	Conclusion on measurement options for radiation hardness characterisation	86
6.4		Radiation hardness at room temperature	86
	6.4.1	Experimental setup	87
	6.4.2	Experimental results	89
6.5		Radiation hardness at liquid helium temperatures	93
	6.5.1	Experimental setup	94
	6.5.2	Experimental results	99

		6.5.3	TCT observations	110
	6.6	Summary		111
7	**Liquid helium chamber measurements**			**113**
	7.1	Experimental setup		113
		7.1.1	Beam lines	113
		7.1.2	Signal readout and data acquisition	113
		7.1.3	Detector prototypes	113
	7.2	Results		115
		7.2.1	Collected charge per MIP	115
		7.2.2	Time response	115
	7.3	Summary		121
8	**Conclusion and outlook**			**123**

CHAPTER 1

Introduction

CERN (European Organisation for Nuclear Research) is a scientific research centre founded in 1954 with the goal of finding answers to the questions of fundamental particle physics. With its 20 member states and many more collaborating states, CERN demonstrates and strengthens the possibility of fruitful international cooperation. Since its existence fundamental discoveries have been made, which more than once were honoured with the Nobel prize. In order to enable those discoveries CERN scientists motivate, use and develop high-end technologies. Furthermore education plays a major role in CERN's philosophy. Many students have the opportunity to spend part of their studies in this supportive environment and prepare for their future scientific career.

CERNs main field of research is particle physics and its flagship is the LHC (Large Hadron Collider), the largest storage ring on Earth. The LHC tunnel has a circumference of approximately 27 km and is located 100 meters below the surface. CERN scientists witness a very fascinating period. Thanks to the reliable LHC performance, the ATLAS and CMS experiments could announce in 2012 that they had observed a particle consistent with the long-sought Higgs boson, in the mass region around 125-126 GeV. Following this new discovery, the Nobel prize in physics 2013 was awarded jointly to François Englert and Peter W. Higgs for their theoretical description of a mechanism helping to understand the origin of mass of subatomic particles.

Further important and successfull CERN technology development fields are for example radiation protection, medical applications like cancer therapy and medical imaging, industrial imaging, electronics, cryogenics, superconductivity and vacuum technology. CERN is hence not only a research centre, but also a technology laboratory.

To keep the LHC beams on their trajectories high field superconducting magnets have been designed. Through regular and irregular beam losses, energy is deposited in the coils of these magnets. The transition from the superconducting state to the normal conducting one is then possible.

A severe machine deterioration can occur with high energy and intensity beams. To ensure safe beam operation and to avoid unwanted damages and quenches from beam losses, different feedback and machine protection systems have been developed. This way operational downtimes can be minimised and a high availability of the LHC can be guaranteed. Those protection systems will trigger a beam dump in the case of a critical event.

The Beam Loss Monitors (BLM) measure the energy deposition of secondary particle showers and trigger a beam dump if those are above a predetermined beam abort threshold. It was found that the current BLM system has a limitation close to the interaction points, where the collision debris can mask dangerous beam losses. The solution investigated in this work is to install particle detectors directly in the cold mass of the magnets and therefore measure the particle showers closer to where the losses happen and closer to the magnet coils needing protection.

In **Chapter 2** an introduction to the LHC and its experiments is given. In addition the most sensitive elements inside the accelerator, being the superconducting magnets, are presented together with the challenge of protecting them from beam losses with the Beam Loss Monitors (BLM). The employed detectors are presented. A short overview of beam loss mechanisms and their impact on accelerator elements is also given.

In **Chapter 3** the motivation for the cryogenic Beam Loss Monitoring (CryoBLM) project is explained and the considered detector technologies for this application together with the finally selected ones are presented. The detectors finally selected for further testing are silicon detectors, diamond detectors and liquid helium chambers. The reasons for their selection are stated.

In **Chapter 4** the charge carrier characteristics at cryogenic temperatures in silicon material, diamond material and in liquid helium are given. They were obtained through laboratory measurements together with literature research. For the detectors operation this is an important factor, as all selected detector technologies are based on ionisation of the atoms and subsequent charge carrier transport within the detector bulk. Therefore knowing the charge carrier properties at cryogenic temperatures will further allow the interpretation and better understanding of the beam test signals from the detectors.

Chapter 5 covers the single particle detection measurements at cryogenic temperatures with silicon and diamond detectors.

Chapter 6 is devoted to the radiation hardness measurements of the tested semiconductor particle detectors at room temperature and at liquid helium temperatures. These measurements allow the comparison of diamond detectors and silicon detectors and enable the advantages and disadvantages for the application as CryoBLMs to be identified.

In **Chapter 7** the liquid helium chamber prototypes are introduced and the results from measurements in beam are presented.

In the **last Chapter** the conclusions of this study are stated and an outlook for the future of the project is given.

CHAPTER 2

The LHC and its BLM system

2.1 LHC (Large Hadron Collider)

The LHC is a circular particle accelerator and hadron collider with a circumference of 27 km. It is designed to collide protons with a momentum of 7 TeV/c at a bunch collision rate of 40 MHz. Its main goal is to investigate physics processes at a new energy frontier. It is made of eight straight sections (IRs), where the experiments and the utility insertions are located and eight arcs, where the dipoles are bending the beam for the circular motion. It is an instrument operating two counter-rotating beams, brought to collision at high energy in order to address open questions in physics. Since November 30th 2009 the LHC is the world's highest energy particle accelerator and storage ring. The twin beams of protons have been accelerated to an energy of 1.18 TeV. With a stable beam circulating at 450 GeV for 10 hours, the LHC furthermore has proven its effective operation as a storage ring. In 2012 the LHC beam momentum for collisions was of 4 TeV/c. Since February 2013 the LHC is stopped for the "Long shutdown 1", in which a major upgrade and repair of the machine was planned. Of high priority was the modification of the high current splices between the superconducting magnets, with the goal to reach nominal beam energies with the LHC.

In the CERN accelerator chain (see Figure 2.1), the LHC is the last element as it is designed for a beam with injection energy of 450 GeV for protons. From the proton or ion sources, LINACs(LINear ACcelerators) increase the particle energies to 50 MeV. Passing through the LEIR (Low Energy Ions Ring) for ions or the PSB (Proton Synchrotron Booster) for protons, the particles reach the PS (Proton Synchrotron), which is successfully running for more than 50 years already. There they reach an energy of 25 GeV and are guided to the SPS (Super Proton Synchrotron), which accelerates the bunches up to the LHC injection energy of 450 GeV. Through the two transfer lines TI2 and TI8 for beam 1 and beam 2 respectively, the particles finally reach the LHC, where they are accelerated to the nominal momentum of 7 TeV/c. For ions the preacceleration is slightly different and in the LHC their nominal momentum is 2.76 TeV/c per nucleon.

The beam is ejected through the dump system located in IR6. This is done regularly under normal operation and also in cases of an unwanted event.

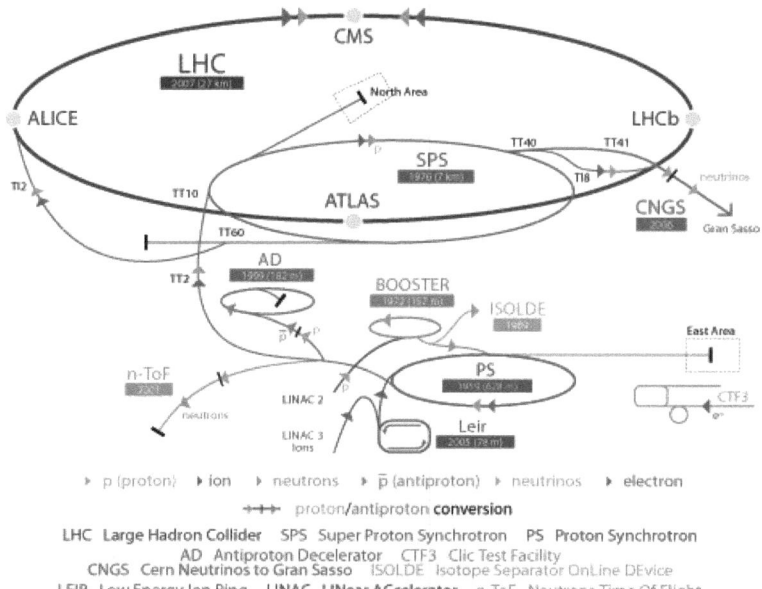

Figure 2.1: Schematic of the CERN accelerator complex. The beams are generated at the LINAC2 (or LINAC3 for Ions) and their energy is gradually accelerated in the BOOSTER (or LEIR for ions), PS and SPS.

2.2 The experiments and their goals

The four big LHC experiments are situated at the interaction points (IP) of the LHC, where the beam trajectories cross each other, the beam size is minimised and head-on collisions are generated.

An important criteria for the quality of the collider is the luminosity, which is related to the collision rate. In the case of colliding beams with a transverse Gaussian particle distribution the luminosity is:

$$L = \frac{N_1 N_2 n_b f_{rev} \gamma_r}{4\pi \epsilon_n \beta^*} F \quad (2.1)$$

where N_1 and N_2 are the number of particles per bunch for beam 1 and 2 respectively, n_b is the number of bunches per beam, f_{rev} is the revolution frequency in the ring, γ_r is the relativistic gamma factor and F is a geometric correction factor due to the crossing angle at the interaction point. ϵ_n is the transverse normalised emittance and β^* is the betatron function at the interaction point. These two parameters describe together the beam size. In order to have a high luminosity the beam intensity has to be maximised (N,n_b), whereas the beam cross section needs to be minimised(ϵ,β). This possibility is limited by the beam-beam effect, where electromagnetic fields are created by the beams themselves. The expected peak

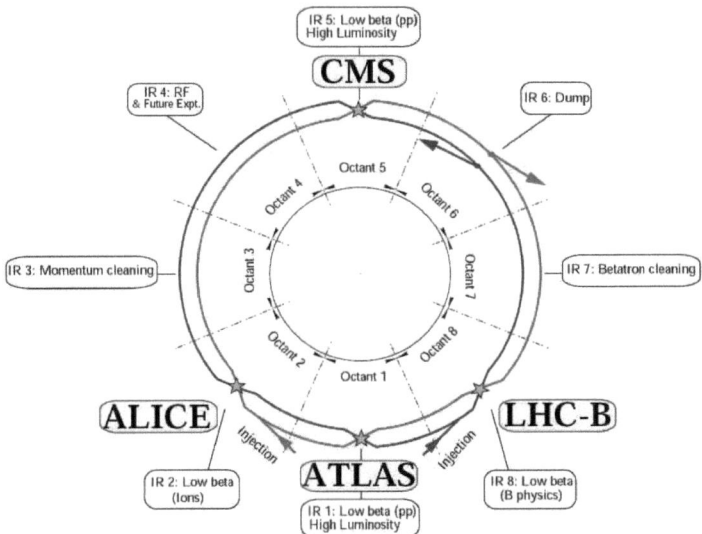

Figure 2.2: LHC Overview with the four main experiments and the accelerator facility areas. The red and blue lines stand for the beam lines and show their relative locations to each other as well as the crossing areas.

luminosity is of 10^{34} cm^{-2}s^{-1} in ATLAS and CMS. The total integrated luminosity delivered by the LHC was of about 23.3 inverse femtobarn (corresponding to $23.3 \cdot 10^{39}$ cm^{-2}) during the year 2012.

ATLAS (A Toroidal LHC ApparatuS) and CMS (Compact Muon Solenoid) are build as detectors for high energy proton-proton collisions. They follow the same general purpose to find the Higgs boson, a particle which existence has been predicted 50 years ago. On July 4th 2012 the ATLAS and CMS experiments held a seminar at CERN. They announced the observation of a particle consistent with the long-sought Higgs boson, in the mass region around 125-126 GeV [1, 2]. The announcements were received with great excitement by the community and many publications in the field followed. Other goals of the ATLAS and CMS experiments are the search for supersymmetry and particles responsible for dark matter. The purpose of those two experiments is the same, but the used technologies are very different.

ALICE (A Large Ion Collider Experiment) is dedicated to the LHC Heavy Ion program. One month per year, heavy ions are accelerated with a luminosity of about 10^{27} cm^{-2}s^{-1}. In the first period high energy Pb-Pb collisions are the goal, which are expected to produce a quark-gluon plasma. The purpose of the experiment is to understand the quark-confinement and the properties of the quark-gluon plasma.

LHCb (LHC beauty) concentrates its researches on CP (Charge Parity) violation in the b-quark system. The question why the Universe seems to be made mainly of matter, but no antimatter is studied.

2.3 Risks from beam for LHC equipment

With an energy of 362 MJ stored in each beam at nominal conditions, the LHC surpasses its antecessor by a factor of 200 (see Figure 2.3). Beam losses can not only negatively affect the stable functionality of LHC machines, but they also can severely damage the equipment. It is therefore important to investigate beam loss origins, their impact on the equipment and design a system able to detect potentially dangerous situations.

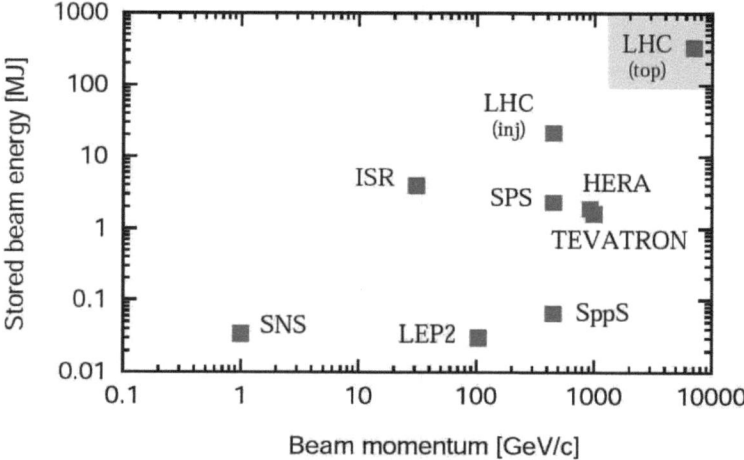

Figure 2.3: Beam energy and beam momentum comparison between existing accelerators. Courtesy of R. Assmann.

2.3.1 Beam loss origins

One can classify the losses into regular and irregular ones. The first type happen even under perfect machine circumstances, while the second one are basically due to malfunctioning. By themselves most of the regular beam losses can be considered as slow compared to the LHC revolution time. They are mainly the root for certain performance limitations. For example they limit the LHC luminosity, can cause beam instabilities or reduce the beam lifetime and provoke emittance growth. Added up and together with machine imperfections, the effects of beam losses need to be understood and can compromise the correct functioning of the LHC.

Examples causing regular losses are [3]:

- Collisions at the interaction points between the counter-wise rotating beams,
- Collimation losses,
- Intra-Beam scattering: small angle Coulomb collisions of two particles from the same bunch,

- Touschek scattering: large angle particles collision, removing particles from the stable radio-frequency bucket and

- Residual gas scattering: protons can collide with the $3 \cdot 10^6$ particles/cm^3 left in the ultrahigh vacuum.

Accidental beam losses are the most dangerous effects. They can be fast and their prediction is impossible. They typically happen in case of operational mistakes or machine failures. The loss rate can then increase very rapidly and endanger machine components. Due to its fast reaction time, the Beam Loss Monitoring system, which will be discussed in detail in the next chapter, is designed to deal with those irregular losses and initiate a beam abort. Later analysis of the post mortem data should then enable to understand the causes of losses, make the right corrections and possibly avoid the error in the future.

Examples for accidental beam losses are: injection losses (those are important and mainly due to mismatches with SPS and transverse injection oscillations leading to losses on IR7 and IR3 collimators), Unidentified Falling Objects (UFOs) [4, 5], kickers malfunctioning (misfiring), operator errors, resonances, beam instabilities and vacuum leaks. UFOs are estimated to be micrometer sized dust particles leading to fast and localised beam losses when interacting with the beam. In 2011 over 16'000 candidate UFO events were recorded and analysed. An example UFO loss will be shown and discussed in section 3.3.1.

2.3.2 Impact of beam losses on accelerator elements

The impact of beam losses on accelerator elements needs to be understood in order to create awareness for sensitive parts, estimate the nature and seriousness of possible damages and design protection systems accordingly. Protons with high energy hitting a target generate a hadronic shower. This interaction leads to energy deposition in the material of the target and an increase of temperature. The consequences can be the transition to the normal conducting state of a superconductor, damages to the LHC equipment and activation of the accelerator elements.

Impact on superconducting magnets

Superconductors are materials that have zero electrical resistivity under certain conditions. The advantage of using superconducting wires in magnets is the possibility to reach high magnetic fields without heat dissipation.

A transition to the normal conducting state has to be avoided. It takes place, if one of the following critical parameters is exceeded:

- temperature T_c,
- current density I_c and
- magnetic field H_c.

This is often summarised in a three dimensional phase diagram as a critical surface.

There are 1232 main dipoles (MB), 392 main quadrupoles (MQ) and a total of more than 4000 corrector magnets installed in the LHC. They are cooled at cryogenic temperatures of 1.9 K and for some of them at 4.5 K. The structural stability in the presence of large electromagnetic forces is achieved by enclosing the magnet coils with a rigid collar made out of austenitic steel. A common magnetic iron yoke surrounds the collars.

A quench, which is the transition from super- to normal conducting state, happens if temperature, magnetic field or current density exceed a critical value. Quenches should be avoided, as they have a negative influence on the magnets lifetime. In addition the coil needs between ten minutes and ten hours (expected time at 7 TeV/c) to recover back into the superconducting state. Dumping the beam before the occurrence of a quench, saves this time for operation. For the main dipole the critical current I_c is above 12960 A, for a magnetic field of 10 T and at a temperature of 1.9 K.

Beam losses can increase the temperature of the superconducting coil and hence induce a quench. The amount of energy a superconducting magnet can absorb before transition to the normal conducting state has been subject to research effort [6, 7, 8, 9, 10]. In the case of the LHC it mainly depends on the beam energy and on the time-scale of the losses. The duration of the losses δt needs to be compared with the time constants τ for heat exchange of the system (see section 2.4.2).

Impact on material

The effect of beam losses on different materials used in the LHC equipment has been studied [11]. The analysed materials are zinc, copper and stainless steel (316L, INCONEL). Cu is of special interest, as it is one of the materials used in the beam screen and in the superconducting cables.

The effect of direct beam losses on Cu plates as targets is shown in figure 2.4. Noteworthy is that no stress related damage, like cracks or twisting, has been observed.

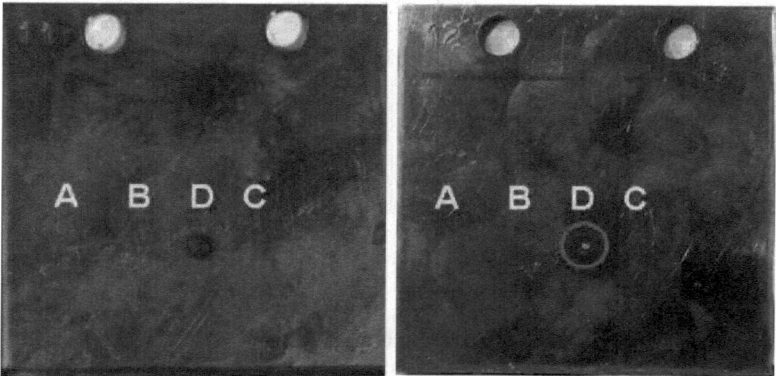

Figure 2.4: Effect of beam irradiation with 450 GeV protons on two Cu plates on different positions. The letter A, B, C and D correspond to tested beam intensities. D is the highest intensity with $7.92 \cdot 10^{12}$ protons and leads to the first damages on the plates. While plate 11 shows just slight discolouration, plate 12 situated closer to the location of maximal energy deposition presents melted material [11].

The understanding of the damage limit is important. The reparation or replacement of damaged elements is time consuming, complicated and expensive. Such damage is a worst case scenario. In the case of typical beam losses, the risk for LHC operation is to affect the correct operation of the superconducting magnet. The limit for quench prevention in terms of critical number of protons is much lower compared to damage protection.

2.3.3 Machine protection

Due to the high beam energy and intensity of the LHC, a sophisticated machine protection system needed to be designed. Figure 2.5 shows the relation between the duration of the losses and its corresponding protection mechanisms.

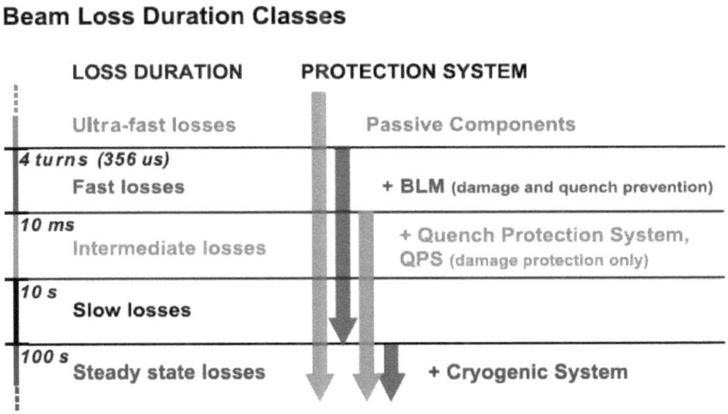

Beam Loss Duration Classes

LOSS DURATION		PROTECTION SYSTEM
	Ultra-fast losses	Passive Components
4 turns (356 us)	Fast losses	+ BLM (damage and quench prevention)
10 ms	Intermediate losses	+ Quench Protection System, QPS (damage protection only)
10 s	Slow losses	
100 s	Steady state losses	+ Cryogenic System

Figure 2.5: Classification of the protection systems according to the loss duration

The LHC protection through collimators is a passive system, enabling the protection from all losses associated with beam instabilities. Collimators allow protection from losses occurring with time scales shorter than the extraction time of the beams. They have one to two moveable jaws that can be positioned in order to define the free path of the beam core. Halo particles on trajectories too large for a continuous circulation are removed so that they can not deposit their energy in sensitive elements any more. This corresponds to a continuous cleaning.

In case of energy deposition in the superconducting magnets, a transition to the normal conducting state is possible. To detect and initiate a protection procedure the Quench Protectin System (QPS) has been developed. It surveys the voltage in the superconducting magnets and if a local quench is detected, it sends a beam abort signal and warms up the whole coil, so that it becomes normal conducting and extracts the current to the external dump resistors. By provoking this transition to the normal conducting state for the whole coil, the energy stored in the magnetic field is dissipated over the entire volume of the superconductor so that major damage can be avoided.

In parallel to the QPS, the Beam Loss Monitoring (BLM) system has been developed, which reacts faster to losses and provides quench and damage protection. The aim of the BLMs is to prevent quenches, while the QPS induces a controlled quench once the magnet stability is endangered. Both therefore avoid damage of the superconducting coils. The QPS is hence an active system, that can trigger a beam dump together with many other LHC protection installations, like the cryogenic installations and the Beam Conditioning Monitoring for ATLAS. Redundancy in the LHC machine protection system is important to guarantee the availability of all systems on one hand and to protect the elements with different principles of the same purpose on the other hand.

2.4 The LHC Beam Loss Monitoring (BLM) system

In order to monitor the beam losses, their evolution and their criticality for surrounding LHC equipment, the Beam Loss Monitoring system has been developed. The main purpose of the BLM system is the quench prevention of superconducting magnets and the damage protection of LHC equipment. The system is designed for fastest losses, with a reaction time (from detection to beam dump initialisation) of about 320 μs, which corresponds to four LHC revolution times. About 4000 detectors have been placed outside the magnet cryostats and measure the radiation of secondary shower particles from lost protons. Should the measured signal exceed a certain threshold, a beam dump is requested to protect sensitive equipment. This threshold needs to be set in a way that a high availability of the LHC for operation is guaranteed. If the threshold is set too low, unnecessary down times are induced by triggering a beam dump, without any potential danger. After the extraction of the beam, the complete LHC machine system needs a certain amount of time to recover until the next beam can be injected and re-ramped. Is the threshold set too high, superconducting coils can get quenched from beam losses without a reaction of the BLM system. This has a negative influence on the lifetime of the magnet and on the operational efficiency, as refilling the machine takes potentially less time than the recovery from a quench.

The placement of the BLMs is critical and needs to be close to where major losses are expected and where the sensitive elements need protection. Typical examples of where losses are expected are the collimation regions and the main quadrupoles.

Not to underestimate is the role of the BLM system as a fast and accurate indicator for the safe operation of the beam. After injection or in case of unwanted happenings, one immediately sees where losses happen and can adjust the settings accordingly. For instance, the BLM system was one of the major tools in the analysis and understanding of the mysterious UFO events (see section 3.3.1).

The BLM system hence is not only a major protection system, but also a necessary diagnostic tool in the daily beam operation.

2.4.1 The BLM Ionisation Chambers

An ionisation chamber is a particle detector, with an active detection medium between electrodes. In an ionisation chamber the applied voltage is of a magnitude that the number of primary created charges is equal to the one collected by the electrodes, without further multiplication or recombination. For gaseous detectors the operational domain corresponds to the plateau region [12] and has the advantage that slight variations of the applied voltage do not influence the response of the detector. This region can be seen in figure 2.6.

The main detectors of the BLM system are ionisation chambers. Figure 2.7 shows a picture of the ionisation chamber and in table 2.1 its most important properties are summarised. They are filled with nitrogen N_2 at 100 mbar overpressure. Parallel aluminum electrodes inside the detector are at a voltage of 1.5 kV. The voltage must be high enough to avoid recombination of positive and negative ions [13]. A charged particle entering the BLM ionises the molecules of the filling gas on their way through the chamber. While the released electrons are accelerated towards the anode, the ions are moving to the cathode. A charge proportional to energy deposited by the particles can be measured.

With the W-factor an estimation of the energy deposited in the detector from the collected charges is possible. This factor is the average amount of energy necessary to create one electron-ion pair.

In order to liberate an electron from its orbit in the molecule a certain energy is needed.

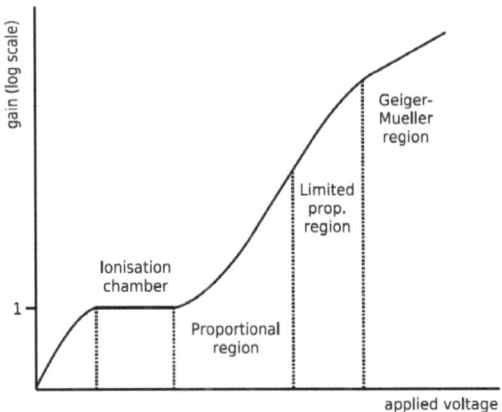

Figure 2.6: The LHC beam loss ionisation chamber.

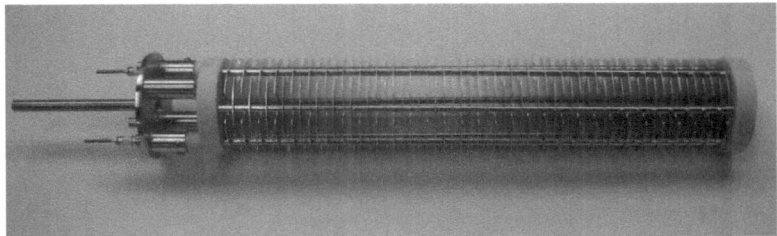

Figure 2.7: The LHC beam loss ionisation chamber.

For N_2 the first ionisation energy is of about 15.58 eV, whereas its W-factor is around 34.8-36.4 eV [14], slightly depending on the type of particle entering the detector. The W-factor is higher than the needed ionisation energy, as additional loss processes are possible, which do not contribute to ionisation.

The response of the LHC BLM ionisation chambers as a function of particle type, its momentum and incident angle is analysed in [15]. Those response functions are shown in figure 2.8.

To cover a higher signal range, secondary emission monitors have been developed and installed at points, where higher losses are expected [16].

2.4.2 Quench preventing thresholds

The threshold defines the limit between safe losses and losses at which equipment is considered to be in danger. In this study the considered equipments are the superconducting magnets and the danger is the transition to a normal conducting state. If the detector threshold is exceeded, the beam will be dumped to ensure machine protection.

An important point that needs to be taken into account for the threshold setting is the operational efficiency: the threshold needs to be set in a way that the magnet is protected

Property	Value
Detector gas	N_2
Length [cm]	50
Diameter [cm]	8.9
Sensitive volume [l]	1.52
Sensitive volume length [cm]	38
Pressure [bar]	1.1
No of electrodes	61
Electrode spacing [mm]	5.75
Electrode thickness [mm]	0.5
Electrode diameter [mm]	75
Standard bias voltage [kV]	1.5

Table 2.1: Properties of the LHC BLM ionisation chamber.

from quenches, but at the same time unnecessary down-times of the LHC are avoided. This down-time is among other things due to the fact, that the dump system needs to be rearmed, the magnet currents have to be trimmed down and the pre-cycling of the magnets has to be done to cancel the remanent fields. If the BLM system triggers a beam dump for innocuous losses, an optimal operation would be impossible. A too high threshold setting on the other hand could have the consequences of quenching or even damaging the superconducting coils before a reaction of the BLM system occurs.

In order to find the optimal threshold, the following parameters have to be determined:

1. the amount of energy that can be deposited in the magnet coil without the induction of a quench (quench level),

2. the amount of energy that is deposited in the superconducting magnets from particle losses for different scenarios and

3. the corresponding signal measured in the BLMs.

The two last points are depending on the energy of the circulating beam. Additionally the first parameter also depends on the time-scale at which the beam is considered to be lost and the last two parameters vary with the loss location and hence with the loss pattern.

The identification of each component enables the estimation of the safety critical BLM thresholds for different time constants, beam energies and loss locations [17, 18, 19]. Those thresholds were determined for worst case scenarios and have been additionally lowered by a safety factor of 3. During 2011 and 2012, with more LHC operation experience and several quench tests with beams, the thresholds were adapted by increasing them for some cases to allow a better availability of the machine and decreasing them for other loss cases as for example for steady state losses.

The effect of different time scales of beam losses will further be introduced through two extreme cases:

- transient losses and
- steady state losses.

Transient losses are faster than any time depending mechanism in the considered situation, while for steady state losses a heat flow needs to be taken into account.

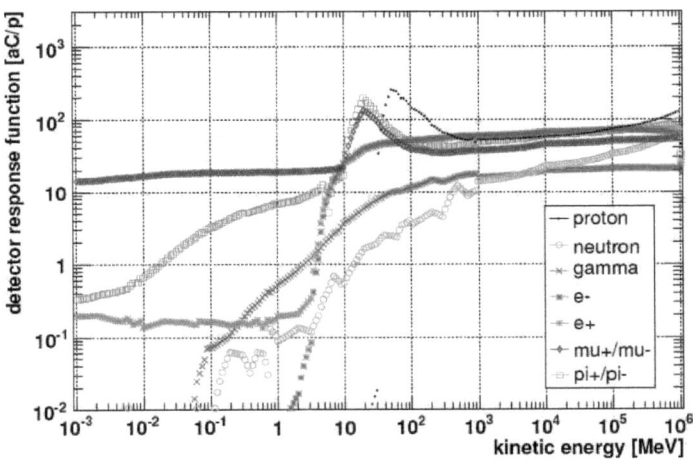

Figure 2.8: BLM response functions for different particle types with an impacting angle of 60 degree relative to the detector axis [15].

Transient Losses

The time scale for this type of loss is short compared to the duration of the heat exchange with the cooling system. Hence no time depending mechanism needs to be taken into account and the parameter of interest is the total radiation dose in Gy.

The following equation defines the threshold $T_{transient}$ for transient processes:

$$T_{transient} = \frac{H_{limit}}{E^D_{max}} Q_{BLM} = N_{critical} Q_{BLM} \qquad (2.2)$$

with:

E^D_{max}: Maximum energy density deposition in the coil of the average lost proton in mJ/cm^3.

H_{limit}: Enthalpy limit of the coil in mJ/cm^3.

Q_{BLM}: Signal in the detector in Gy.

$N_{critical}$: Number of critical protons.

Steady state losses

For transient losses a central parameter is the energy deposition inside the superconducting coil, while for steady state losses the criteria is the power density. Furthermore a heat transition to the cryogenic cooling system needs to be taken into account. Current studies suggest that the bottlenecks for the heat transfer are the helium channels in the insulation of the superconducting cables.

The following equation defines the threshold $T_{steadystate}$ for steady state losses:

$$T_{steadystate} = \frac{w_q}{E_{D,cable}} Q_{BLM} = \dot{n}_q Q_{BLM} \qquad (2.3)$$

where $\dot{n}_q$ is the quench limit in protons per second, w_q is the quench margin defined by the heat transmission capability, $E_{D,cable}$ is the energy density deposition in the cable and Q_{BLM}

is the signal inside the chamber. The dimension of the thresholds for steady processes is given in μGy/s, as time depending mechanisms need to be taken into account.

2.4.3 The limits of the present BLM system

The impact of collision products on the accelerator complex downstream of the interaction points was studied and optimal protection strategies were implemented [20]. These proved to be effective during the 2011 and 2012 LHC operation. Nevertheless the beam loss monitors are sensitive to the collision debris, which has consequences on the protection ability of the system.

The inner triplets are four superconducting magnets (quadrupoles Q1, Q2A, Q2B and Q3) installed on both sides of every Interaction Point (IP). These magnets were designed and implemented to squeeze the beams before the experiments, in order to maximise the collision efficiency (see section 2.2 and equation 2.1) and therefore also the physics discovery potential. The triplet magnets have high magnetic fields to perform the pronounced squeeze of the beams. These strong magnetic fields increase their risk for transition from the superconducting state to the normal conducting one, as the critical temperature decreases with higher fields. Seventeen Beam Loss Monitors are installed along the triplet magnets with the purpose to prevent their quench.

Particle shower simulations in the triplet magnets

Particle simulations were performed with the Monte Carlo code FLUKA to estimate the deposited energy in the coils, together with the signals in the BLMs for different scenarios and time scale of the losses [19].

The simulation geometry contains all relevant components of the LHC insertion region up to 300 meters on the right side of IP1 (ATLAS). In the simulation also magnetic fields, LHC tunnel and the ATLAS cavern were implemented. The part of the geometry containing the triplet magnets can be seen in figure 2.9. For the BLMs around the triplets a detailed FLUKA model was used, containing gas, electrodes, chamber walls and the electronic components [21].

The proton-proton collisions were simulated with DPMJET III [22] for a center of mass energy of 14 TeV. The results of the collision simulation were normalised to a nominal luminosity of 10^{34} s^{-1} cm^{-2}, assuming a proton-proton interaction cross section of 80 mbarn.

The simulated loss scenario is based on wrong collimator settings and leads to a maximum energy deposition in the middle of the triplet magnet (Q2B).

In the detailed comparison of the collision debris with the loss scenario of wrong collimator settings, it became clear that the current BLM setup can not fully protect the magnets in the steady state case. Due to the proximity of the interaction point, a differentiation between signals from dangerous accidental losses and from the continuous collision debris is difficult.

The steady state results from the simulations can be seen in figure 2.10, where the effect of the high radiation fields due to collision debris is visible. The most important result of the figure can be seen in the comparison of the lower two curves. Both correspond to the signals in the BLMs (where each small cross on the curve is the signal from one BLM, connected with a line to guide the eye). The blue curve corresponds to the signal from simulated collision debris and can be considered like a background, while the red one is the signal of interest from a dangerous loss scenario. The violet line stands for the proposed BLM beam abort threshold, which loses its purpose because of the large debris background. The two upper curves show the dose deposited in the magnet coils, where again blue is used for debris and red for losses

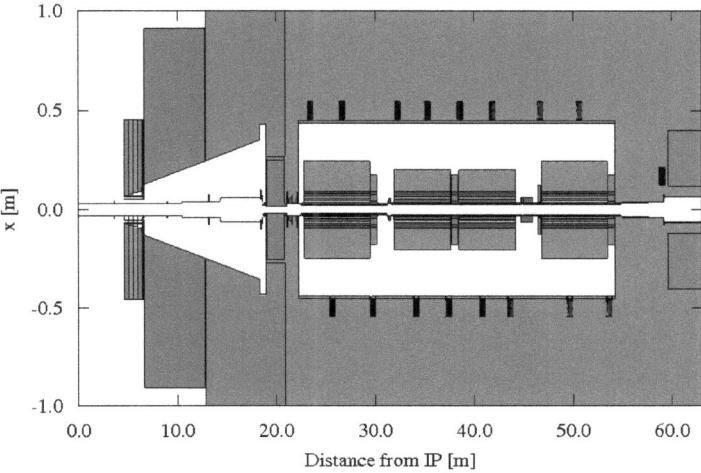

Figure 2.9: FLUKA simulation geometry of the triplet magnets [19]. Note: the scales on the axes are different.

from wrong collimator settings. The maximum of the red curve corresponds to the estimated quench level of the superconducting coil. In the presented case, a clear distinction between debris and dangerous losses is visible.

In summary the figure 2.10 shows: collision debris are of no danger for the magnets (the implemented protection strategies are successful), but the signal from the debris is exceeding the threshold for safe beam operation. The reason for this is that the collision debris consist to a major part of pions with a momentum distribution peaking at 500 GeV/c. When these particles arrive in the magnet, they are strongly deviated, leading to a large impact angle. The resulting particle showers are hence approximately transverse. On the other hand the proton impact angle from losses due to wrong collimator settings is in the order of mrad. The resulting particle showers in this case lead to less signal in the BLMs (relatively to the energy deposition in the coils) compared to the collision debris.

With the present configuration of the installed Beam Loss Monitors (BLM) in the triplet region, the ability to distinguish between energy deposition in the coil due to beam loss is limited by the debris, masking the beam loss signal. The problem can be overcome by implementing "topological thresholds", for which the beam abort threshold is not set for each detector separately, but certain patterns and loss profiles of several monitors can be taken into account. Such a conditional threshold could be done through electronic logic and would increase the precision of the system for a beam dump. The procedure was estimated to delay the reaction time by about 20 ns. This delay is acceptable for the BLM system, but the setting of conditional thresholds depending on loss location, loss scenario and beam energy is error prone.

The simulations have been done for nominal beam intensity only and increased the awareness for future situations in the High Luminosity LHC. The limitation in the triplet magnets

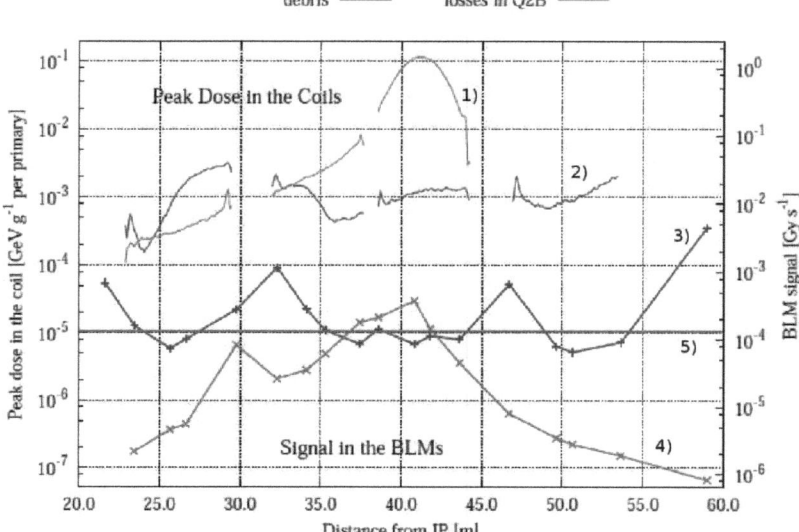

Figure 2.10: The curves represent the following: 1) Deposited dose in the coils from losses inside Q2B, 2) Deposited dose in the coils from the debris, 3) BLM signal from debris (one cross for each BLM), 4) BLM signal from dangerous losses inside Q2B, 5) Proposed BLM beam abort threshold to protect from dangerous losses. One can see that the debris can mask the signal from a dangerous loss and that the proposed threshold is not adequate.

is the best studied case, but with beams of higher energy (compared to 2012) and intensity resulting in higher luminosity, the current BLM limitations will become even more critical and are expected to cause limitations at several locations in the LHC, e. g. close to collimators and close to the beam transfer lines from the SPS to the LHC (injection regions). The BLM experience of the last years during several quench tests showed already that the BLM signals at quench can vary. Two quench tests were compared in detail [23]. In one case with diluted losses, the signal was 30 % above the expected quench threshold and no magnet quench occurred, while in the second case the signal at quench was 3 % above the threshold.

A long term solution needs to be found allowing a more pronounced raise of the actual energy deposition measurement inside the superconducting coils of the magnets above the background level from collision debris.

The primary goal of the simulations was to estimate the beam abort thresholds. The results lead to a dedicated research and development topic to investigate the further presented solution for the limits of the current BLM system.

CHAPTER 3

The cryogenic BLM project

At the triplet magnets, close to the interaction regions of the LHC, the current Beam Loss Monitoring system is sensitive to the particle showers resulting from the collision of the two beams. In the future, with beams of higher energy and intensity resulting in higher luminosity, distinguishing between these interaction products and possible quench-provoking beam losses from the primary proton beams will be challenging. The situation is expected to be similar in the cleaning insertions, where collimators absorb particles that are too far from the nominal trajectory. Investigations were therefore done to optimise the system by locating the beam loss detectors as close as possible to the superconducting coils of the triplet magnets. These Cryogenic Beam Loss Monitors (CryoBLMs) have hence to be located inside the cold mass in superfluid helium at 1.9 K. Different detector technologies have been considered and will further be described in this Chapter.

3.1 CryoBLM as solution for the present BLM limitation

A solution based on placing radiation detectors (CryoBLMs) inside the cold mass close to the coils is investigated. The advantage would be that the dose measured by the detector would more precisely correspond to the dose deposited in the superconducting coil. This proposition is based on bringing the situation from the current BLM signal seen in figure 2.10, closer to the dose in the coil, where a better distinction between debris and losses is possible.

The new possible detector placement is depicted in figure 3.1. In the cross section of the triplet magnets one can see four holes of 10 cm diameter, allowing the flow of liquid helium through the iron yoke. Three of the holes are free for the installation of detectors.

3.1.1 Requirements for the cryogenic BLM application

The main challenges for this cryogenic detector inside the cold mass are:

- the low temperature of 1.9 K (superfluid helium),
- the integrated dose of about 2 MGy in 20 years,
- the reliable operation in a magnetic field of 2 T,

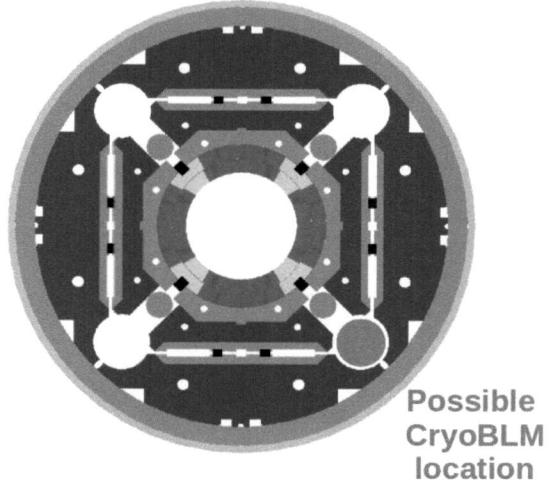 

Figure 3.1: Cross section of the triplet magnet (courtesy of Paolo Ferracin) with the current BLM placement in yellow and the possible location for a cryogenic BLM in green. The superconducting coils are in red, while the iron yoke is in blue colour. The design of the triplet magnets for the HL-LHC is still in the conceptual phase and changes of the exact structure of the iron yoke are possible. The presence and availability of one of the four holes is certain.

- the mechanical resistance to a fast pressure rise up from 1.1 to about 20 bar, in the case of the quench of a magnet,
- the time response faster than 1 ms.

This is a completely new and demanding set of criteria that has never been required or investigated in such a form before. A certain knowledge about particle detector characteristics is available around 2 K [24], due to applications in space and other experiments in liquid helium. Unknown is the combination of the cold with the total radiation dose of 2 MGy.

The CryoBLM specifications for the PhD research topic were set to a time response faster than 1 ms. This value is due to the fact that the detectors should be in future not only installed in the triplet magnets, but also at further locations in the high luminosity LHC. These locations require a maximum time response of 1 ms.

There are two main concepts for the detector readout:

- Direct Current (DC) measurements and
- particle counting mode.

The DC measurements are favoured for the Beam Loss Monitoring application, because they are sensitive to the amplitude of the signal and because the LHC losses have a DC like behaviour, building up turn by turn. The downside of particle counting is that events resulting in different signal amplitudes (corresponding to a different deposited dose) will have the same weight in the decision whether a loss was dangerous or not. Different amplitudes are possible as particles of different types and energies are expected. In addition particles arriving at the same time within about 10 ns, which is highly probable due to the LHC bunch structure, will count as one event only. In counting mode the correlation between deposited dose and number of events is therefore weaker compared to DC measurements of the losses. The possible advantages of particle counting are further discussed in section 3.2 and Chapter 5.

Once the detectors have been installed into the cold mass of the LHC magnets, no access will be possible. Therefore the detectors need to be available, reliable and stable for 20 years. In the case of the selection of detectors sensitive to radiation damage, the evolution of their properties must be well known and controlled.

3.2 Considered detector technologies

The initial strategy was to search for already existing experience and technologies in literature and to discuss the topic with different experts. This lead to a collection of different detector types that were analysed for their potential as future CryoBLM. In the following part, the operating principle of dismissed detector technologies will be shortly introduced and the reasons for them not being further tested and considered for the project will be given.

3.2.1 Gaseous Helium3 ionisation chamber

^{4}He is by far the most abundant isotope of helium. The other stable isotope ^{3}He is only of about 0.1 ppm of natural helium, but it can also be manufactured. It is a product from the tritium beta decay, requiring the high neutron flux of a nuclear reactor.

Helium3 ionisation chambers are often used in nuclear industry because of their high sensitivity to thermal neutrons and their simple and reliable construction and operation. The high absorption cross section for neutrons is due to the nuclear reaction:

$$n +^3 He \rightarrow ^3 H +^1 H + 0.764 MeV \qquad (3.1)$$

in which the electrically neutral helium-3 atom is converted into two charged particles that can directly be detected with an ionisation chamber.

The boiling point of helium-3 under atmospheric pressure is at a temperature of 3.19 K [25]. Under a pressure of 0.02 atm, helium-3 is in the gaseous state at 1.9 K. This would allow the operation of a gaseous ionisation chamber as CryoBLM, which is impossible with any other detection material at 1.9 K. In the case of a warm up to room temperature of the LHC magnets, the pressure within the ionisation chamber would increase. The estimated pressure of the container would go up to 3.27 atm. This can be easily mechanically withstood.

The company Saint Gobain [26] produces custom designed Helium-3 proportional counters, typically used for neutron detection. Several discussions with Saint-Gobain representatives and further investigations indicated that Helium3 ionisation chambers seem to have convenient properties for the CryoBLM application, but Helium3 is rare and expensive (3000 €/liter of gas in February 2011). The production of ^{3}He decreased sharply with the international drive to stop the production of nuclear weapons after the cold war. Based on the actual shortage and the unclear future situation, it was decided to not invest further resources in the study of this material.

3.2.2 Thermal equilibrium calorimeters

A thermal equilibrium calorimeter consists of three main components:

- the thermal mass for the absorption of the energy from particles,

- the thermometer to measure the temperature increase ΔT and
- a weak thermal link between the thermal mass and the heat sink at a constant temperature T_0.

These 3 components and the operating principle of a thermal equilibrium calorimeter can be seen in figure 3.2. ΔT is proportional to the energy deposition of the particles and is measured with the thermometer. τ is the cooling time constant of the thermometer, giving a measure for the time needed to decay back to T_0.

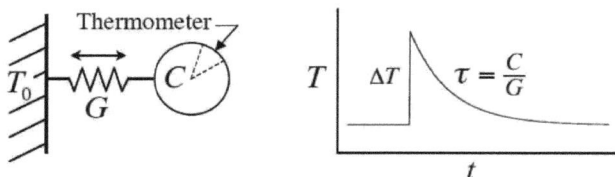

Figure 3.2: Principle of operation of an ideal thermal equilibrium calorimeter [24]. C is the heat capacity, G is the thermal conductivity and T_0 is the temperature of the heat sink.

In thermal equilibrium calorimeters, there are no requirements for efficient charge transport, meaning that impurities can be tolerated. In addition thermal equilibrium detectors enable a very good energy resolution (3 eV FWHM at 6 keV), because of the very large number of phonons generated with correspondingly small statistical fluctuations [24].

The implementation of a cryogenic beam loss monitor for the LHC has been considered in 1996 already [27]. The investigated solution was based on a thermal equilibrium calorimeter. In the attempt, a carbon resistance acted as thermometer, thermally coupled to two copper blocks. In case of losses, particle showers deposited heat into the copper, leading to a resistance decrease that was measured. The resistance dependence on temperature changes were well characterised. The microcalorimeter design is shown in figure 3.3. This setup was first used in the laboratory for precise measurements and calibration of all relevant components and later in beam tests.

These measurements in beam made it possible to correlate the resistance decrease with the number of particles. The found dependence can be seen in figure 3.4, where the measured temperature variation is plotted as a function of the number of protons, proving the functionality and linearity of the system.

The microcalorimeter was not used for the LHC, because of the long cooling time constant of 3 s. This time constant is limiting the time resolution of the system and is not acceptable for safety critical LHC beam loss measurements.

3.2.3 Liquid helium scintillation light

A particle traversing liquid helium excites the helium atoms. The excited atom assumes the ground state through the emission of scintillation light and phosphorescence. The scintillation light is in the Extrem Ultra Violet (EUV or XUV), centered around a wavelength of 80 nm [28].

The scintillation light can be detected:

- indirectly, with a wavelength shifting material to obtain longer wavelengths, which are measurable with classical photo-multipliers and

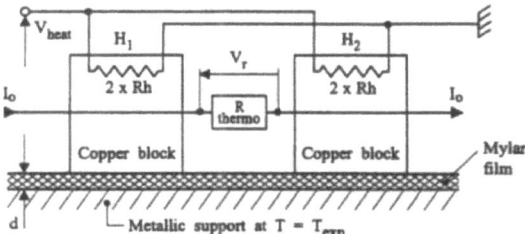

Figure 3.3: Microcalorimeter design used in the laboratory (with H1 and H2 as heaters for energy deposition) and in the experiment with beam [27].

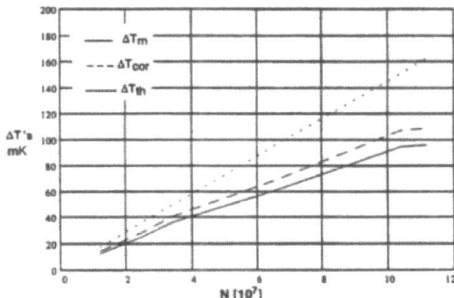

Figure 3.4: Results from the microcalorimeter experiment with beam [27]. ΔT_m is the measured dependence (lowest line), ΔT_{th} is the theoretical dependence (highest line) and ΔT_{cor} is the corrected measured temperature increase.

- directly through the use of semiconductor detectors.

The downsides of the indirect technique is the complex assembly and structure of the wavelength shifters together with the shifting related signal loss. The EUV to visible conversion efficiency is typically of about 30 %. In addition wavelength shifting optical fibers reduce the time response. The strongest argument against their use is that they are far from radiation hard enough for the application in the LHC magnets over 20 years. Due to their complexity, low efficiency and non-radiation hardness, wavelength shifters were therefore excluded.

The other option to detect the EUV photons is to use semiconductor particle detectors. The photon energy E can be calculated through:

$$E = h\nu = \frac{hc}{\lambda} \quad (3.2)$$

where h is the Planck constant, c is the speed of light, ν is the frequency and λ is the wavelength. The resulting photon energy at a wavelength of 80 nm is of 15 eV. The energy needed to create an electron-hole pair is for silicon material of 3.6 eV and for diamond material of 13 eV. EUV photons are therefore able to create charges inside solid-state detectors. The charge generation in the detector would therefore be of 1 to 3 electron/hole pairs per EUV photon. As the range of the photons in silicon and diamond material is very short, the surfaces of the detectors, the metallisation and the doping need to be adapted to allow the detection of these photons.

The passage of a minimum ionising particle (MIP) generates 24 000 electrons in a 300 μm thick silicon detector. The indirect detection of the MIP using the produced scintillation light with a solid-state detector corresponds to the MIP exciting the helium atoms, which would emit EUV light equally distributed in all directions. From these photons only a small amount would reach the detector to create electron-hole pairs with small efficiency. Should future research refocus in this direction, the charge carrier properties and the radiation hardness of the solid-state detectors in liquid helium need to be known. These points are treated in the following chapters of this work and can be used for future investigations of scintillation light detection with semiconductors. The indirect detection of particle showers through helium scintillation light was not considered as beneficial compared to the direct detection of the particles. The method was hence rejected.

3.2.4 Transition edge sensors

Transition Edge Sensors (TES) are superconducting films operated in the temperature range between normal and superconducting state. They measure an energy deposition by the increase of the resistance of the superconducting film, as shown in figure 3.5. Their main advantage is their high sensitivity, allowing an excellent energy resolution of 1.4 eV FWHM at 100 mK for X-ray applications [29]. The required transition temperature can be reached using bilayers and combinations of superconductors. In addition, arrays with superconductors of different sensitivity can be arranged to increase the dynamic range of the measurement. For the BLM application a large dynamic range of 9 orders of magnitude is required (the current BLM system can measure from pA to mA), which means the implementation of arrays with different combinations of superconductors, being a highly complex system.

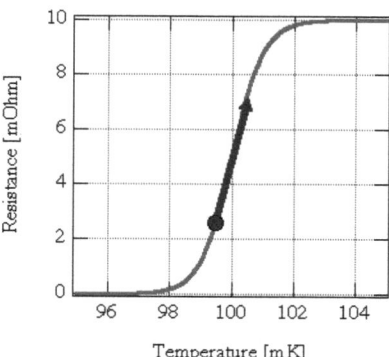

Figure 3.5: Transition Edge Sensor operating principle [29].

In addition, the TES have the tendency for instability due to Joule heating and possible fluctuations in the liquid bath temperature. The operating principle of the TES is furthermore similar to the already existing Quench Protection System (QPS) of the LHC, where the voltage between two points of a superconducting cable of the magnet is measured. If the voltage exceeds a predefined threshold, a beam dump is triggered and a controlled quench of the magnet is induced through heaters.

Overall the TES could therefore not be further considered for LHC's cryogenic beam loss monitoring.

3.2.5 Scintillators

Scintillators are materials emitting light under radiation. The scintillation process can be described as the conversion of an incident particle or energetic photon into a number of photons with much lower energy. These photons can further be detected using photomultiplier tubes.

The Crystal Clear Collaboration and Saint Gobin [26] were contacted to get an estimation of the feasibility of using scintillators as CryoBLMs. It was confirmed that there is no direct answer and that the problem is complex. There is a lack of information on the behaviour of the scintillators under the specific conditions.

From measurements during a "frozen spin polarized target" experiment, low efficiency of the scintillators in liquid helium could be observed [30]. The scintillators were exchanged by silicon detectors and the experiment could be resumed with success.

Furthermore the optical fibres for the light transport from the scintillator to the outside of the cryostat where the photo-multiplier would be, are not radiation hard enough. Especially considering that preliminary experiments done at the Frauenhofer Institut showed a tendency of the optical fibers to have rather worse radiation hardness with decreasing temperatures.

The investigations in the direction of scintillating crystals promised little success, therefore other detector technologies were favoured.

3.3 Selected detector technologies

The considered detection technologies are the main topic of this work. Silicon detectors, diamond detectors and a liquid helium ionisation chamber were selected and tested in the laboratory and with beam. The following discussion will show why these technologies have been chosen. The operating principle of all the selected detector technologies is based on the generation of free charge carriers in the detection material and subsequent collection of these charges. The charge transport phenomena are therefore of major importance for all three detector types and a full chapter (Chapter 4) is dedicated to the understanding of the charge carrier properties.

3.3.1 Solid-state particle detectors

The focus in this section will lie on the reasons for choosing the following detectors, together with their application relevant properties.

Silicon detectors

Very good and detailed introductions to the exact working principles of semiconductor detectors (and especially silicon detectors) can be found in the following literature [31, 32, 33].

The available knowledge and the existing experience in using silicon detectors as BLMs and in using them at cryogenic temperatures, were two of the main reasons for their selection.

At DESY in Hamburg, silicon devices were in use as beam loss monitors in counting mode for the HERA-proton-ring [34]. PIN-diodes were under discussion also for the LHC beam loss monitoring system before the decision was made to use ionisation chambers.

The CERN RD50 collaboration is investigating the radiation hardness of silicon devices. A meeting was organised with its spokesperson Michael Moll to get his opinion on the usability of silicon detectors as future CryoBLM.

The CERN RD39 collaboration is specialised in the research of silicon radiation hardness at cryogenic temperatures. It is within their collaboration that the Lazarus effect (charge collection recovery of silicon at lower temperatures), the current injection mode (CID) and the increased radiation hardness of silicon detectors at temperatures below 230 K could be investigated [35, 36, 37]. The CryoBLM project was presented to the RD39 collaboration in 2010, leading to further common effort in the characterisation of silicon down to liquid helium temperatures.

The CID mode corresponds to an external current injection, allowing the manipulation of the electric field distribution within the detector in such a way that full depletion is expected to be given at any fluence. The CID results from the RD39 collaboration looked promising and motivated further investigations. A description of the CID mode working principle can be found in section 4.3.3.

A further major advantage of the silicon detector operation at cryogenic temperatures is that the leakage current decreases significantly, because the thermal energy is too low to ionise the donor atoms. The current-voltage (I-V) characteristics for a pn-silicon diode is given through the Shockley equation:

$$I = I_0 \left(e^{\frac{qV}{k_B T}} - 1 \right) \tag{3.3}$$

where I is the diode current, I_0 is the reverse bias saturation current, V is the applied voltage, q the electron charge and T the absolute temperature. In this equation, a negative voltage corresponds to a reverse bias operation.

Silicon detectors were already used in at CERN in 1976 at liquid helium temperatures [30]. A scattering experiment at the CERN Proton Synchrotron (PS) used a "frozen spin" polarized target. For the experiment silicon surface barrier diodes were operated at 1 K, replacing scintillators that showed low efficiency. The pulses in silicon were generated from 5 GeV/c negative pions traversing the target. The diodes showed good performance at 1 K. The noise was strongly reduced and the FWHM of the pulse went from 50 ns at room temperature to 20 ns at 1 K. The readout electric circuit can be seen in figure 3.6 and is very similar to the one used during the CryoBLM experiments for single particle detection. One side of the detector is connected to ground and to the cable shielding, while the other side of the detector is used for signal readout and voltage application at the same time. No radiation damage was observed after 15 weeks of operation with 10^5 particles/s (total about 10^{12} particles). In addition there was no detectable effect on the operation of the diodes from a 0.1 T magnetic field.

In another experiment silicon detectors were immersed in liquid helium and irradiated with $1.2 \cdot 10^{14}$ neutrons/cm^{-2} of 1 MeV [41]. The team found a reduction of the destructive effects compared to room temperature. Several observations could be done:

- the trap concentration went from $5 \cdot 10^{12}$ cm^{-3} to 10^{14} cm^{-3},
- the full depletion voltage was reached at 20 V already for irradiated diodes,
- for temperatures below 150 K, the leakage current was below the system noise of about 1 pA,

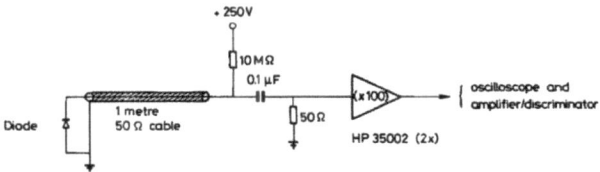

Figure 3.6: Electric circuit used in the experiment for the readout of a silicon diode [30].

- the forward current at 4.2 K is of only 10 pA for applied bias voltages smaller than 30 V and

They further observed a charge collection efficiency recovery at low temperatures, which they estimated to be due to filled and therefore passivated defects. The downside of this effect was that it lead to an increased polarisation of the detector and might therefore cause a disappearing signal [41].

These results and discussions motivated to further investigate the silicon properties in liquid helium and to find out if it might be the technology of choice for the cryogenic BLM application.

Diamond detectors

The application of diamond material as particle detector is relatively new compared to silicon detectors. The progress in growing high quality synthetic diamond in the last years, improved their characteristics as particle detectors. Detailed information about the diamond detector technology can be found in the following literature [42, 43, 44].

The first operation of diamond detectors as Beam Condition Monitors was at the BaBar experiment [45]. The detectors successfully replaced the ageing silicon PIN diodes in the horizontal plane, where the radiation was most intense in the positron-electron collider.

Polycristalline Chemical Vapour Deposition (CVD) diamond BLMs have been installed in the LHC in the cleaning region IR7, where collimators are close to the beam line and therefore higher losses can be observed [46]. CVD is the most frequently used growth method for diamond detectors. These diamond BLM detectors are not yet included in the beam interlock system and can hence not trigger a beam dump, but the analysis of their signals during critical losses, lead to unprecedented insights. The diamond detectors are also not connected to the BLM electronics, but a different read-out with oscilloscope is used. Data is only saved when a certain trigger level is exceeded. The time resolution of the measurements is of 2 ns, while the minimum integration time of the BLM ionisation chamber electronics is of 40 μs. In the upper plot of figure 3.7, one can see the signals from a diamond BLM in the case of an increasing loss due to an UFO event with a final beam abort. The beam dump was triggered by the ionisation chambers of the current BLM system. The rise time of the losses until the dump was of about 600 μs, corresponding to 7 LHC turns. All plots in figure 3.7 are from the same UFO event. The lower two plots are a zoomed region of the first one and enable to see the analysis potential of the diamond monitors. Not only the filling scheme is visible with its train of bunches, but also a bunch by bunch resolution is possible, allowing to distinguish the 36 bunches with 50 ns spacing. Such a resolution is not needed for protection purposes, but it is a great tool for offline analysis, feedback systems and further understanding of the beam dynamics. These successful measurements together with the future possibilities, additionally motivated to investigate the use of diamond detectors as CryoBLMs.

A further motivation for the investigation of diamond materials were the results of the CERN RD42 collaboration, which is specialised in the investigation and development of radiation hard diamond detector material [48]. The results of their measurements with respect to radiation hardness at room temperature were promising. This is especially true since the fabrication of high quality diamond detectors, which was motivated by the RD42 collaboration.

The advantages of diamond detectors compared to silicon devices at room temperature are:

- the very low leakage current even after irradiation,
- the radiation hardness, because a larger kinetic energy of the incoming particles is required for the displacement of an atom from the crystal lattice,
- the higher charge mobility,
- a more stable operation under temperature variation,
- the insensitivity to visible light and
- the intrinsically lower noise level.

These properties are valid at room temperature and no comparison exists down to liquid helium temperatures. The main properties of silicon and diamond detectors are depicted in table 3.1.

The successful use of diamond detectors as research beam loss monitors together with their advantages compared to silicon devices made clear that not only silicon detectors should be tested, but also the potential of diamond sensors for the CryoBLM application needed to be investigated.

Remark for operation in counting mode

A further remark with respect to the measurements shown in figure 3.7 concerns the possible BLM operation in counting mode. In this read-out mode, each bunch is expected to lead to only one count in case of this example loss, because all particles generated by one bunch arrive in a time shorter than the time resolution of the semiconductor detectors with 300 μm thickness. It would be an additional development to measure in parallel the increase of the signal amplitude with time. The counting mode would detect an increased number of bunches over a certain count-threshold within specific time windows. This is not equivalent to an increase of the amplitude of the losses and strongly depends on the bunch structure. The missing information about the signal amplitude makes a reliable correlation between the number of counts and the deposited energy in the magnet coils difficult.

3.3.2 Liquid helium ionisation chamber

The idea of using the liquid helium itself as detection medium came directly after the rejection of the use of helium-3 gas. The main advantage is that noble liquids are stable and insensitive to radiation damage.

In comparison to solid-state detectors, liquids can fill up the entire wanted experiment volume, leaving no clearance. The size of liquid detectors can be more easily adapted to the needs of the experiment. Liquids furthermore saturate in general at a higher level of energy

	Silicon	Diamond
Atomic number	5.43	3.57
Atomic mass	14	6
Mass density [g cm^{-3}]	2.329 [49]	3.515 [49]
Melting point [K]	1685 [49]	4100 [49]
RT Band gap [eV]	1.124 [49]	5.48 [49]
LHe Band gap [eV]	1.17 [50]	5.45 [50]
Breakdown field [V/μm]	30	1000
Relative dielectric constant	11.9 [49]	5.7 [49]
Energy to create eh-pair [eV]	3.6 [51]	13 [52]
Intrinsic resistivity [Ω cm]	$2.3 \cdot 10^5$	$> 10^{15}$
Detector capacitance [pF]	8...10	2
Thermal expansion coefficient [10^{-6} K^{-1}]	2.59 [49]	0.9 [55]
Thermal conductivity [W cm^{-1} K^{-1}]	1.4	21 [55]
Ionisation loss from MIP [MeVcm^2/g]	1.67 [44]	1.76 [44]
Mean charge per MIP [fC/100 μm]	1.4 [44]	0.62 [44]
Displacement energy [eV]	25 [56]	43 [57]

Table 3.1: Comparison table between diamond and silicon detectors. The capacitance is given for the following detector dimensions: 0.3x5x5 mm^3 for Si and 0.5x5x5 mm^3 for diamond.

deposition than for example crystal detectors [58]. They have a higher potential for superior energy resolution and linearity [59]. In comparison with gas detectors, liquids produce typically three orders of magnitude more electron-ion pairs per unit length, making liquid detectors intrinsically more precise.

The usual disadvantages of liquid detectors are that they need purifiers, cryogenics, leak-proof containers with precise temperature control. In the case of the CryoBLM project this is not a disadvantage, but a given condition within the cold mass of the magnet. The criteria for purity during the liquefaction process of helium are demanding. After rigorous filtering, an impurity check is performed and should the impurity level be above 5 ppm, the gas is rejected for liquefaction. For the helium gas with impurity levels below the threshold, one further stage of gas cleaning and filtering is carried out. In addition to this process the properties of the superfluid helium together with the temperature of 1.9 K reduce the probability of impurities leading to an influence of the charge generation and collection. The major disadvantage in of the liquid helium is the expected slow charge collection, because of the small charge mobilities in comparison to other detectors. This will further be treated in the section 4.5 on charge properties in liquid helium.

For neutrino detection, the use of liquid helium as detection medium is investigated in the USA. The aim of the project is to develop a new cryogenic neutrino detector, an "electron bubble chamber" to detect low energy neutrinos from the sun. The team managed to perform the experimental proofs of the operating principle and gave a further outlook in the following paper [60].

During conferences, the CryoBLM project and first results were presented. The attention for the project increased and in fact the idea to use liquid helium as detection medium for beam loss monitoring already existed from Russian colleagues in 1992 [61]. They also point out the slow charge collection as one of the major disadvantages. They never had the chance to build or test such a detector, but in 2012 a meeting at CERN was organised and the first liquid helium chamber prototype together with the first results of measurements in beam were presented to them.

3.4 Summary

Simulations have shown that the capability of the current BLM system to fully protect the LHC superconducting coils from quenches is limited by the collision debris. This motivated the investigation of the possibility of installing detectors into the cold mass of the superconducting magnets.

From all considered detector technologies, silicon detectors, diamond detectors and a liquid helium based particle detection are the most promising for this application.

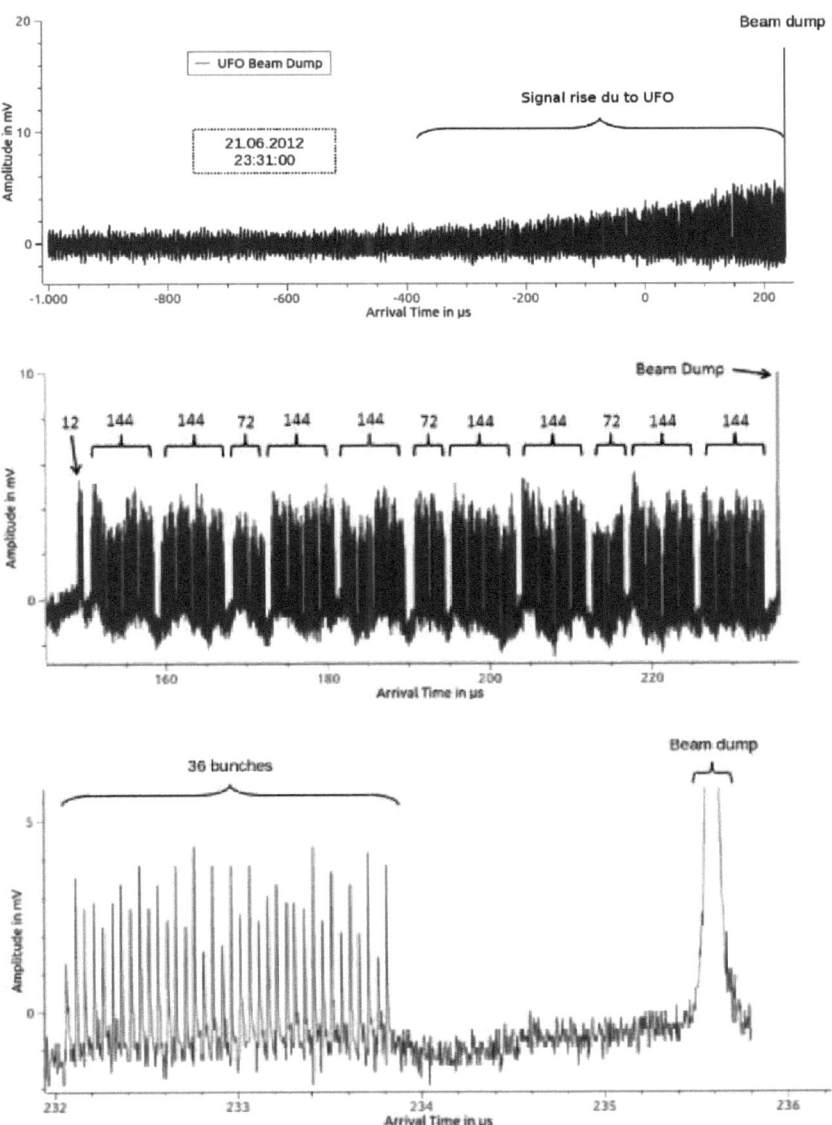

Figure 3.7: Diamond BLM signals from one UFO event [47]. On the x-axis is the time in μs, while on the y-axis is the signal amplitude in mV. The first plot shows the entire event, while the next ones are zooms, depicting the analysis potential of the detector setup. Injection scheme and bunch structures are visible.

CHAPTER 4

Charge carrier characteristics at cryogenic temperatures

This chapter covers the properties of charge carriers in the selected detector material at cryogenic temperatures. These properties need to be understood, as the detector's particle detection characteristics strongly depend on an efficient charge carrier transport.

First the experimental Transient Current Technique (TCT) to obtain the charge carrier properties in semiconductor material is introduced. Then the results for silicon material and single crystal CVD diamond material are presented. The last section is dedicated to the description of the unusual charge carrier structures that arise in liquid helium and to estimations of important liquid helium chamber characteristics.

The results of this chapter will allow to interpret and understand the later following low intensity beam test measurements. The measured charge carrier mobilities and drift velocities will furthermore be used in the modelling of the degradation of the detector signals during the high intensity beam tests.

4.1 The transient current technique

The Transient Current Technique (TCT) is a method to investigate and analyse the main detector properties, mostly focused on the charge transportation phenomena inside the bulk of the detector.

Charges are generated on the surface of one side of the detector through a laser light or alpha particles. The penetration depth of alpha particles is of only about 10 μm and also the laser wavelength can be chosen in such a way, that the light only penetrates the surface of the detector. Depending on the electric field, electrons or holes drift through the detector bulk and their signal can be recorded. A simplified scheme of the TCT method is shown in figure 4.1. The electric circuit used for this measurement is the same as shown earlier in figure 3.6.

The shape of the resulting pulses at different temperatures and different voltages enables further analysis of the detector charge characteristics. The properties that can be analysed with the TCT method are:

- electric field distribution in the detector,
- charge drift velocities and mobilities,

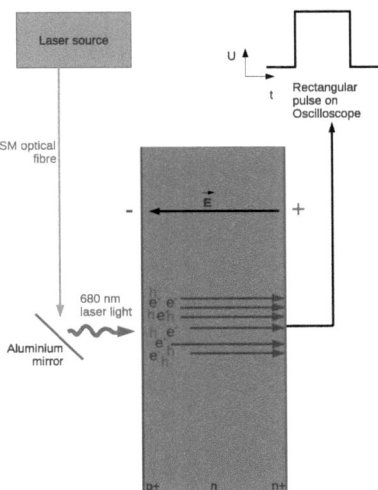

Figure 4.1: Simplified operation principle of the transient current technique. The charges (electron - hole pairs) were generated by the 678 nm laser light (or alpha particles) directly at the surface of the detector and depending on the electric field, electrons or holes drift through the detector bulk. In the drawing only one laser mirror is shown for simplification. In fact two mirrors are installed, allowing to measure the hole and the electron drift separately. The charge drift was visualised and recorded with the oscilloscope. In the ideal case the shape of the charge drift is rectangular. The real case is often different due to for example charge trapping, detrapping, space charge and electric field non-uniformities. The shape can hence be used for a detailed analysis of the detector properties, especially for the comparison of different detectors, with different voltages, at different temperatures and different irradiation levels.

- minimum electric field for full charge collection (full depletion voltage),
- space charge sign,
- charge trapping and
- charge detrapping.

TCT measurements do not allow to conclude on the charge collection efficiency of the detector for other particle types and energies, especially for MIPs. This is due to:

- the dominance of surface effects like parasitic (not bulk related) effects from the doped regions in silicon and
- an extremely dense charge cloud (10^{21} cm^{-3}) at the impact location of the alpha particle (this density is not given in the case of laser illumination, due to its low intensity and the standard deviation of the Gaussian beam, which was estimated to be around 1.2 mm),

It can not be guaranteed, that the amplitudes of the signals in the case of laser measurements can be perfectly compared for different temperatures. The laser light transmission from the laser head to the detector surface is expected to be temperature dependent. This is due to e. g. possible temperature dependent losses in the used optical fibers or the effect of the thermal contractions on the optical connectors and the installed aluminium mirrors (to direct the laser light on the detector), leading to slight laser light alignment variations.

4.2 Cryogenic basics

The final application of the cryogenic detectors will be in the liquid helium environment, at 1.9 K. In order to make tests and characterise the usability of different detectors, a cryostat was necessary. This cryo-cooler was filled with liquid helium, in which the detectors were immersed. The boiling point of helium at atmospheric pressure is at a temperature of 4.32 K. This temperature was lowered by reducing the pressure of the gas space above the ^{4}He bath. By pumping away the helium gas molecules, the temperature of the entire bath was reduced and controlled by maintaining a constant pressure in the cryostat. The heat capacity of liquid helium at 4.2 K is 5.3 J/K for one gram of liquid, while the one of copper is only 10^{-4} J/K/g [62]. It is therefore with a good approximation that only the heat capacity of the liquid helium needs to be taken into account, while the heat capacity of the apparatus can be neglected.

The cooling by mass evaporation can be described in the following way [63]:

$$L(T) \cdot dm = m \cdot c(T) \cdot dT \qquad (4.1)$$

where L is the latent heat, T the temperature, m the mass and c the heat capacity. The latent heat of vaporisation of liquid helium at 4.2 K is of 20.8 J/g. The relation between the reduction of mass from m_i to m and of the temperature from T_i to T can be written as:

$$ln(\frac{m}{m_i}) = \int_{T_i}^{T} \frac{c(T)}{L(T)} \cdot dT \qquad (4.2)$$

To cool liquid ^{4}He from 4.2 K to 1.9 K about 50 % of the initial helium mass needs to be evaporated and pumped away. A major part of this mass is needed for the phase transition from normal fluidity to superfluidity.

4.3 Charge carrier characteristics in silicon material

The experimental setup for measurements down to liquid helium temperatures is further described. The results from the TCT measurements with laser are shown and allow conclusions concerning the charge carrier characteristics in silicon material. The used laser wavelengths were 678 nm (red region) for the TCT measurements and 1060 nm (infra-red region) for the MIP simulation.

4.3.1 Experimental setup

The silicon diode in its TCT holder and the cryostat insert are shown in figure 4.2. The connection with the outside of the cryostat was done through optical fibre feedthroughs and SMA feedthroughs on the top plate of the cryostat. A tube through the cryostat top plate allowed to move the silicon detector from the ground of the cryostat up to the top of the cryostat. On the bottom of the cryostat was the liquid helium, while the area above the copper shields was close to room temperature.

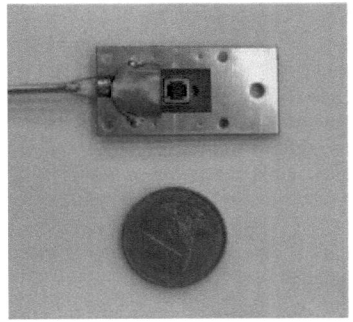
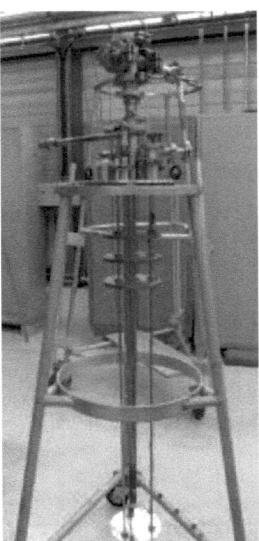

Figure 4.2: The left picture shows the silicon diode in its holder (from the Ioffe Institute in St. Petersburg) used for TCT laser tests down to liquid helium temperatures. Further structures were added to hold aluminium mirrors and optical fibres, guiding the light onto the detector sides. The right picture shows the insert for the cryostat. The detector within the TCT module was on the bottom of the construction.

The TCT measurement readout was done using a current amplifier (20 dB and 40 dB, 2 GHz, 180 ps rise time) from CIVIDEC to amplify the signal coming from the silicon samples. The acquisition was done with a broad-band 3 GHz LeCroy oscilloscope. The picosecond laser light was generated by a PiLas Digital Control Unit (EIG1000D) and an optical head for 678 nm. This laser wavelength generated charges on the surface of the silicon diode. The laser instrument allowed to vary the laser tune and the repetition frequency. The laser tune was proportional to the output power and was set in percent, where 100 % corresponded to the maximum laser output power. The laser repetition frequency defines the repetition rate at which the laser pulses were generated.

A TCT module held the silicon diode and enabled the installation of two single mode optical fibres (with 9.25 µm core diameter and 125 µm cladding diameter), one for each detector side. The laser light is guided from the single mode optical fibre through two aluminium mirrors on the detector sides, as shown in figure 4.1 (with one mirror only for simplification). Electrically one side of the silicon was connected to ground and cable shielding, while the other detector side was used to apply voltage and read signal at the same time. This was achieved through an electric circuit as shown in figure 3.6.

The silicon detector was a 10 kΩcm p^+-n-n^+ device with Al metallisation. The thickness of the sample was 270 µm and its active area 3x3 mm^2. The contact doping was done with boron and phosphorus. The boron doped p^+-layer had a thickness of 1000 Å and a boron density of about 10^{18} mm^{-3}. The phosphorus doped n^+-layer had a thickness of 2 µm with a phosphorus density of 10^{18} mm^{-3}, while the n-type silicon bulk had a positive effective space charge of $3.6 \cdot 10^{11}$ mm^{-3}.

The temperature, the pressure and the liquid helium level were monitored during the measurements. A CERNOX temperature sensor was placed at the height of the detector.

4.3.2 Reverse bias measurements

Laser parameter control

The effects of the laser power and frequency variation on the detector signals are shown in figure 4.3. Each of the shown pulses in this section was made out of an average of a minimum of 100 signals and the representation of the average pulses was done using histograms, in which the bars in y-direction of the pulses show the standard deviation of the measurements, while the bars in x-direction correspond to the width of the histogram bins. The pulse shapes for different laser repetition frequencies were less subject to changes compared to the power variation, where the signal amplitude increased linearly with the tune.

The effect of the laser repetition frequency and the power variation on the width measurement is shown in figure 4.4. While the laser repetition rate had an impact on the FWHM of the pulses, the influence of the tune on the FWHM of the signal was not significant. The standard deviation of the noise was of 2.5 mV during the measurements and its influence on the calculation of the total width of the signal increased for lower laser powers.

During the measurements, the laser tune and its repetition frequency was kept constant at all time at a value of 70 % and 500 Hz. The rise times of the signals was limited by the used electronics and was measured using a pulser as input to the amplifier. The obtained lower rise time limit was of 180 ps, which was small enough to not lead to any influences on the signal shape interpretations.

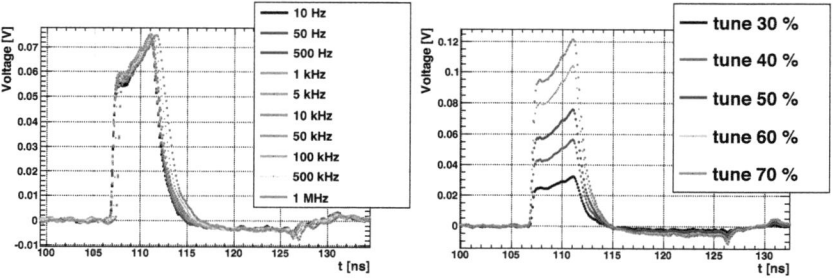

Figure 4.3: Variation of the TCT pulse shape at 5.5 K under laser repetition frequency (left plot) and laser tune (right plot) modifications. After the pulses, reflections are visible from the rising and the falling edge of the signal with a delay of 20 ns, corresponding to the 2 m cable length between detector and amplifier. The frequency repetition rate had an impact on the FWHM. The reflection coefficient is negative and is due to a slight impedance mismatch between detector and amplifier input.

Charge carrier drift pulses

Figure 4.5 shows the signal pulses at constant electric field for holes travelling through the detector bulk at different temperatures. To obtain the hole drift properties, the laser illumination is performed on the n$^+$ contact of the silicon detector. The pulse shape changes

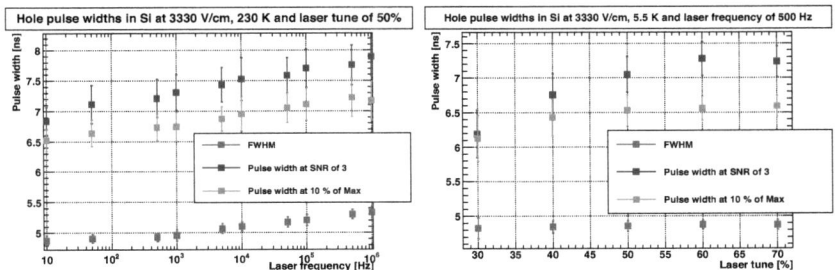

Figure 4.4: Variation of the pulse width at 5.5 K under laser repetition frequency (left plot) and tune (right plot) modifications. The x-axis of the frequency variation plot is in logarithmic scale.

were significant and the detector property variations with temperature were clearly visible. All measured pulse shapes were different from the ideal rectangle.

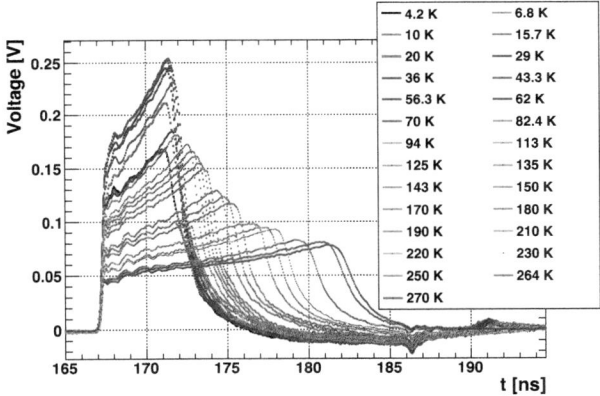

Figure 4.5: Average hole drift pulses at different temperatures under a constant electric field of 3330 V/cm.

The n-type bulk of the tested silicon detector had a positive effective space charge of $3.6 \cdot 10^{11}$ mm^{-3}, due to which the hole drift was accelerated within the detector bulk. This is visible in the top part of the pulse shapes, as the increase of the pulse amplitude for the hole drift from the start of the signal to its end. The slope is larger at low temperatures, because the drift time is smaller, while the number of space charges stays constant. For irradiated silicon detectors the situation would be different, as the induced damages act as charge traps and modify the effective space charge.

In figure 4.6 the signal pulses at a constant temperature of 4.2 K for holes travelling through the detector bulk at different electrical fields can be seen. A full depletion of the silicon detector was given at only 10 V already.

Figure 4.7 shows the signal pulses at constant electric field for electrons travelling through the detector bulk at different temperatures. To obtain the electron drift properties, the laser

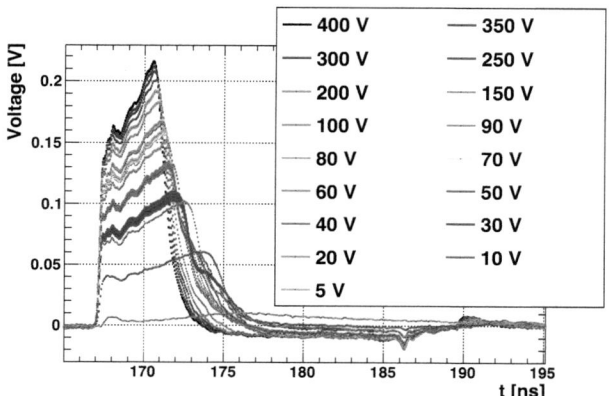

Figure 4.6: Average hole drift pulses at 4.2 K for different applied voltages. The curves at 30 V and 40 V with larger error bars correspond to a moment with slight instabilities of the light output of the laser head.

illumination is performed on the p^+ contact of the silicon detector. Again the pulse shape changes are significant. While the hole drift in figure 4.5 was accelerated within the detector, the electron drift was decelerated within the bulk, due to the positive space charge. This is visible in the decrease of the pulse amplitude for the electron drift from the start of the signal to its end.

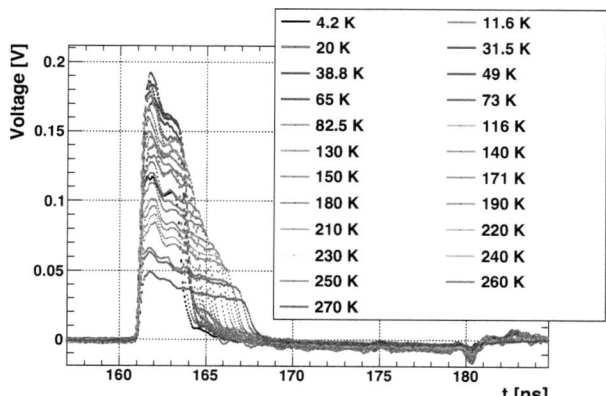

Figure 4.7: Average electron drift pulses at different temperatures under a constant electric field of 3330 V/cm.

In figure 4.8 the signal pulses at a constant temperature of 4.2 K for electrons travelling through the detector bulk at different electric fields are shown.

Figures 4.6 and 4.8 depict that for low voltages the shape was very different from the ideal rectangle. The signal possessed a very long tail and the shape was more comparable to

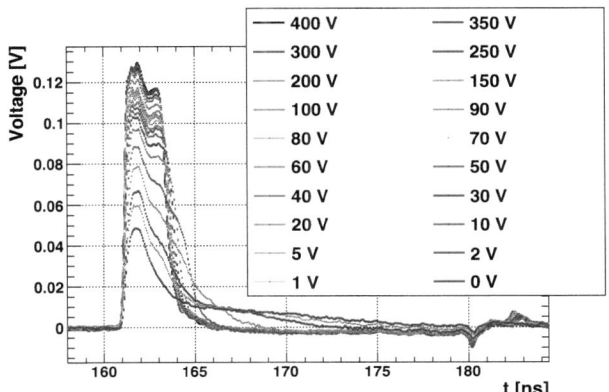

Figure 4.8: Average electron drift pulses at 4.2 K for different applied voltages.

a triangle for some cases. This was due to the fact that the electric field was not high enough to fully deplete the detector and cover its entire thickness. The charge drift was then partly due to diffusion.

Collected charge from the laser light

The collected charge Q generated by the laser light was estimated by integrating the signal. The obtained charge collection efficiency (CCE) is depicted in figures 4.9, for electric field and temperature variation. As already mentioned before, the observed changes for different temperatures were most probably not only due to changes in the detector alone, but also due to laser light transmission dependence on temperature. The temperature variations of the TCT amplitudes can therefore not be considered for conclusions regarding the detector's CCE.

The electric field dependence at constant temperature on the other hand was measured under stable laser light transmission. The measurement of the charge versus the applied electric field E allowed the estimation of the full depletion voltage of the silicon detector sample.

Charge carrier drift velocities

The pulse widths correspond to the time the charges need to drift from one detector surface to the other. The obtained drift times from the measurements are shown in figure 4.10. A pulse width decrease with temperature decrease and electric field increase was observed.

The charge drift velocity v_{drift} was estimated from the drift times t_{drift} through the detector bulk thickness d:

$$v_{drift} = \frac{d}{t_{drift}} \qquad (4.3)$$

The obtained drift velocities in silicon are depicted in the figure 4.11, showing their increase with reduced temperature.

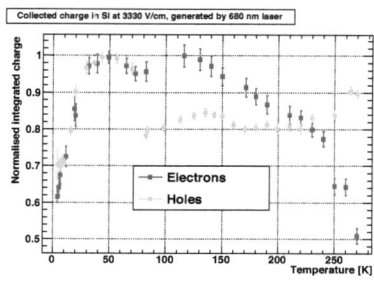
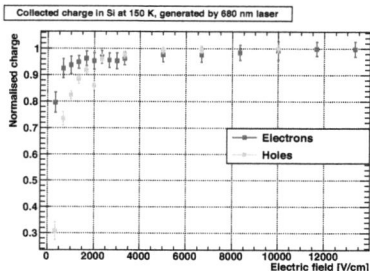

Figure 4.9: Dependence of the TCT charge collection efficiency on the temperature (left plot) and on the electric field (right plot) for hole and electron drift pulses. The variations with temperature were most likely not only due to property changes of the detector material, but also due to laser light transmission variations.

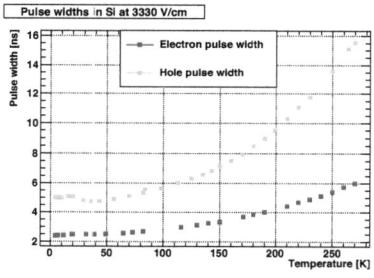
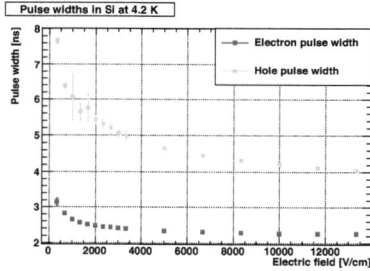

Figure 4.10: Electron and hole drift times dependence on the temperature (left plot) and on electric field (right plot).

Charge carrier drift mobilities and saturation velocities

The right graph of figure 4.11 illustrates that the linear relation between the drift velocity and the electric field E is not given for all field strengths. Introducing the low field mobility μ_0 and the saturation velocity v_{sat}, the relation can be described as [65]:

$$v_{drift} = \frac{\mu_0 E}{1 + \frac{\mu_0 E}{v_{sat}}} \qquad (4.4)$$

where the low field mobility μ_0 corresponds to the linear slope for low electric fields and the saturation velocity v_{sat} is the velocity towards which the curve asymptotically tends for high electric fields. The calculated low field mobilities and saturation velocities versus temperature are depicted in figures 4.12 and 4.13 respectively. The calculated saturation velocities are in agreement with earlier publications [53, 54], while the charge carrier mobilities in silicon material are strongly impurity dependent. The calculated electron saturation velocity at 4.2 K was $(1.20 \pm 0.01) \cdot 10^5$ m/s, while in [54] it is of $1.3 \cdot 10^5$ m/s. The estimated hole saturation velocity at 4.2 K was $(6.67 \pm 0.03) \cdot 10^4$ m/s, but it was not well determined and not reached with the applied fields in this work. This is due to the "anomalous behaviour of the curve of holes at low temperatures (T < 30 K)" [54]. Saturation of the hole drift velocity had not been

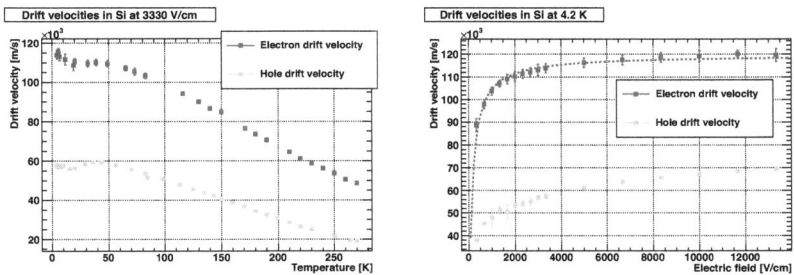

Figure 4.11: Electron and hole drift velocity dependence on the temperature (left plot) and on electric field (right plot).

observed experimentally and is expected to occur above fields of $2 \cdot 10^5$ V/cm with a value of 10^5 m/s [54]. The low field mobility dependence on the impurity density was measured and explained for a temperature of 300 K [54]. The detector tested in this work had a phosphorus density of $3.6 \cdot 10^{11}$ mm^{-3} in the n-type bulk, The calculated low field electron mobility for this density was $(8.27 \pm 0.66) \cdot 10^4$ cm^{2}V^{-1}s^{-1} and of $1.3 \cdot 10^5$ cm^{2}V^{-1}s^{-1} in [66] at 8 K with an impurity density of $4 \cdot 10^{10}$ mm^{-3}. The calculated hole low field mobility was $(2.37 \pm 0.6) \cdot 10^4$ cm^{2}V^{-1}s^{-1} at 4.2 K and of $1.1 \cdot 10^5$ cm^{2}V^{-1}s^{-1} in [67] at 11 K with an impurity density of 10^{11} mm^{-3}. The results for the tested detector in this work are summarised in table 4.1.

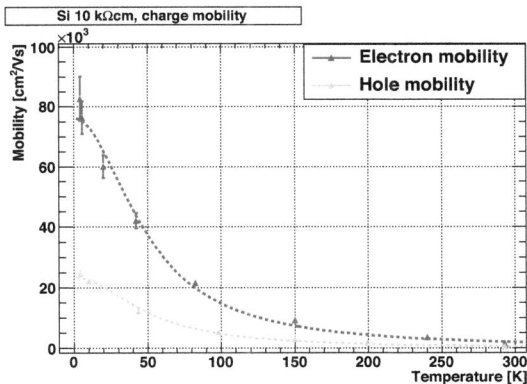

Figure 4.12: Dependence of the electron and hole mobilities on the temperature.

The TCT method was therefore successfully used to determine the basic properties of the charge transport in silicon material down to liquid helium temperatures. It is a powerful method and the obtained results on drift velocities and mobilities agree with earlier publications [54], in which other methods were used for the charge mobility measurements.

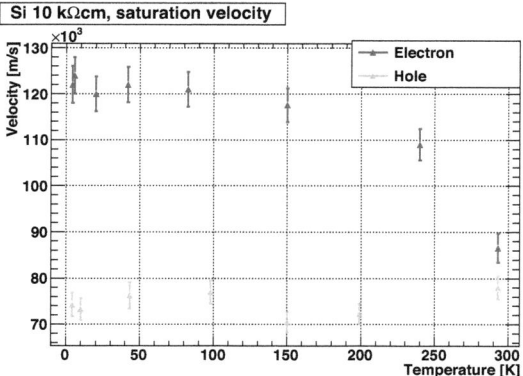

Figure 4.13: Dependence of the electron and hole saturation drift velocities on the temperature.

4.3.3 Silicon detector forward bias operation

CID mode Introduction

Under the reverse bias operation of the silicon devices no full depletion can be reached after a certain fluence. This means that parts of the detector bulk do not see an electric field. Charges generated in these regions can therefore only be collected by diffusion, which is much slower and increases the trapping probability. The Current Injected Detector (CID) corresponds to an external current injection, allowing the manipulation of the electric field distribution within the detector in such a way that full depletion is expected at any fluence.

The injected charges in CID mode are used to fill deep levels from irradiation defects by trapping. The cooling of the detectors enables to reach a non equilibrium state between trapping and detrapping time constants in such a way that trapped charges are not detrapped again [38]. The trapping time constant τ_t is:

$$\tau_t = \frac{1}{\sigma v_{th}(T) N_t} \tag{4.5}$$

where σ is the trapping cross section of the deep level, $v_{th}(T)$ is the thermal velocity, T is the absolut temperature and N_t is the concentration of the deep level.

The detrapping time constant τ_d is:

$$\tau_d = \frac{1}{\sigma v_{th}(T) N_v e^{-\frac{E_t}{k_B T}}} \tag{4.6}$$

where N_v is the effective density of states in the valence band, E_t is the activation energy of the deep level and k_B is the Boltzmann constant. With decreasing temperatures the detrapping time constant increases exponentially. Especially at liquid helium temperatures, no noticeable detrapping - even of shallow levels - is therefore expected to occur.

The electric field manipulation is hence obtained through the manipulation of the space charge density in the detector by trapping of injected charge carriers. The electric field as a function of the depth inside the detector is shown in figure 4.14 for different carrier injection currents. For low current densities, the applied voltage does not fully deplete the diode, meaning that the electric field does not span the entire bulk of the detector. With increased current densities the electric field reaches further into the bulk. The ideal electric field distribution is found with the optimum current injection density j_{opt} of 16 nA/cm² for this detector under the measurement conditions. The practical problem is that this optimum current injection density depends on the applied voltage, the temperature and the actual fluence. The challenge would therefore be that the ideal injected current has to be adapted for different temperatures, fluences and bias voltages. This is of little practical use during experiments and difficult to optimise continuously. In addition first current injection was performed through laser illumination, leading to additional disadvantages in the feasibility of the operation modus, as the light source, the optical feedthroughs and the optical fibres must be installed, stable and radiation hard.

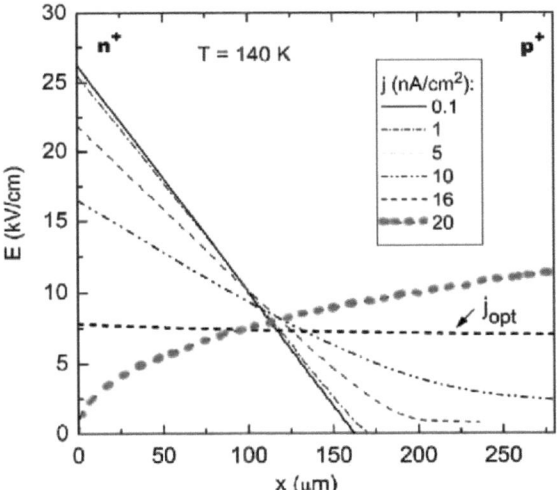

Figure 4.14: Dependence of the electric field distribution on the injection current density in an irradiated silicon detector [39]. The electric field in CID operation is highlighted in red.

A simple solution to these difficulties is the CID operation under self stabilised current injection [39], which corresponds to operating an irradiated p⁺-n-n⁺ silicon detector in cryogenic temperatures under forward bias. This is shown as the curve in figure 4.14 with an injected current density of 20 nA/cm², which is above the optimum current. This electric field has the disadvantage of being non-uniform, but the advantages are that it spans the detector bulk and is estimated to be constant for any fluence. The calculations and investigations of the electric field optimisation by carrier injection leading to this CID operation proposal have been performed in [40].

The resulting electric field E in CID mode with self stabilised current injection is:

$$E(x) = \sqrt{\frac{2V^2}{d^3} \cdot x} \qquad (4.7)$$

where x is the coordinate along the detector bulk thickness d, with $x = 0$ corresponding to the n$^+$ contact. V is the applied voltage. This means the self-stabilised electric field is zero at $x = 0$ and increases with the square root of the distance from the injection contact (n$^+$).

The corresponding injected current density J is:

$$J = \epsilon\epsilon_0\mu V^2 \frac{p}{p_t d^3} \quad (4.8)$$

where ϵ and ϵ_0 are the permittivity of silicon and the vacuum respectively, μ is the charge mobility and p and p_t are the concentrations of free holes and trapped holes respectively. The forward current is therefore proportional to the second power of the applied voltage in the self-stabilised mode.

First CID property measurements have been performed with silicon detectors irradiated to different fluences with neutrons at room temperature. These detectors were cooled down after irradiation to temperatures between 180 and 280 K to perform experiments on the CID operation mode [39]. The observations with increased fluence were that the noise and the forward current decrease. In addition the Charge Collection Efficiency (CCE) was improved compared to reverse bias.

Unfortunately from the publications no direct conclusions for the CryoBLM application are possible, as the detectors were irradiated at room temperature with neutrons and the lowest tested temperature is of 180 K. In addition the CCE decreased towards lower temperatures. Should this trend continue until 1.9 K, the signal could completely disappear.

Based on publications, the overall CID operation properties at liquid helium temperatures were unpredictable and needed further testing, which was successfully performed in this work.

Measurement observations

For the CID operation to work between 180 and 280 K, a minimum of irradiation defects is needed to manipulate the space charge density and therefore the electric field distribution within the detector bulk. Due to the decrease of the CCE in CID mode with decreasing temperatures the question was if this operation modus will work at liquid helium temperatures on irradiated detectors. The unexpected result was that the CID is operational even for non-irradiated silicon detectors below 20 K. The obtained pulses under forward bias operation are shown in figure 4.15.

The forward current measurements shown in figure 4.16 lead to a resistance of the silicon device of 2.041 ± 0.004 MΩ and a resistivity of 6.45 ± 0.01 MΩcm. This large current of 98 μA at 200 V forward bias is a disadvantage for the use of such a detector as CryoBLM with DC read-out. In addition this current would lead to power dissipation inside the cold mass of the LHC magnets, which rather endangers than protects the correct machine functioning.

The forward bias modus is nevertheless of major interest. Not only could these measurements prove that the CID mode works at liquid helium temperatures, but it is also predicted that the forward operation is more radiation hard than the reverse bias modus. The future CryoBLM detector could hence be a forward biased silicon detector operated in counting mode or even in DC, as the high forward current is supposed to disappear after a certain fluence. This will be shown in section 6.5.2.

4.3.4 Infra-red light pulses

The laser light with a wavelength of 1060 nm generates charges over the entire thickness of the silicon detector, this is because the absorption length for this laser light is much larger than

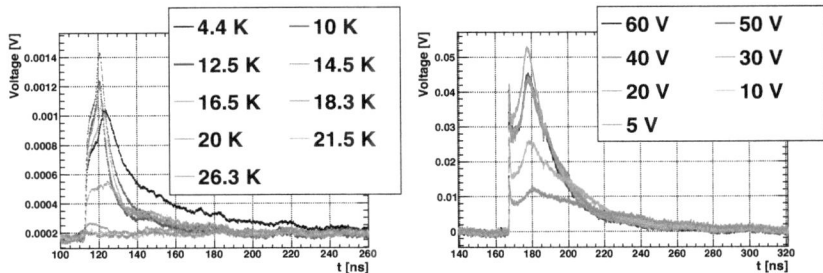

Figure 4.15: Pulses of the non-irradiated silicon detector under forward bias and laser light injection on the n^+ contact side. In the left plot the applied voltage was of 200 V and pulses are shown for different temperatures. The maximum signal amplitude was reached at a temperature of 10 K. In the right plot the signals were measured at different voltages and at a constant temperature of 4.2 K. The maximum forward signal amplitude was obtained for a forward applied voltage of 30 V. A major difference between the two sets of measurements, is that for the temperature variation no amplifier was used. This is visible in the height of the pulses, but also in disappearance of parasitic shapes in the pulse, like the very first peak of all pulses in the voltage variation. This peak was not due to charge transport phenomena within the detector, but due to amplifier characteristics.

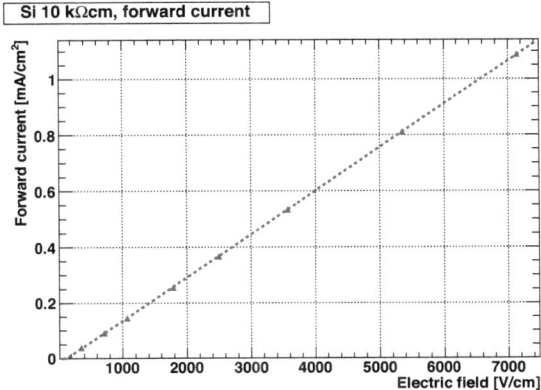

Figure 4.16: Forward current of the non-irradiated silicon detector. The forward bias current variation is below 0.48 % in the temperature range between 2 and 27 K.

the detector thickness. It can therefore simulate the pulse shape of traversing MIPs. When charges are generated over the entire thickness of the detector, the resulting pulse shape will be triangular. The pulse shapes at room temperature are shown in figure 4.17. The drift of the electrons can be distinguished from the one of the slower holes. This was again observed during the beam tests and is shown in figure 5.17. As the electrons have a faster drift through the bulk, their falling edge from the triangular pulse shape is shorter and more pronounced, while the falling slope of the holes is visible after the one of the electrons.

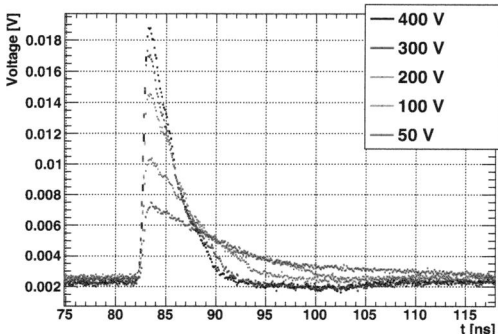

Figure 4.17: Infra-red light pulses of the non-irradiated silicon detector at room temperature under different applied voltages.

With 1050 nm laser light, the signal disappears at low temperatures because the absorption of infra-red light is strongly temperature dependent. This has already been observed and measured in the RD39 collaboration [35] with the use of their cryogenic TCT setup [68]. These measurements did hence not allow to conclude on the MIP pulse shapes at liquid helium temperatures, but an experiment in beam has been designed (Chapter 5) and single MIPs were measured.

4.4 Charge carrier characteristics in diamond material

The TCT for single crystal CVD diamond was done with an Americium alpha source ^{241}Am, which has been tested and approved for the use in vacuum and at cryogenic temperatures. Similar to the case of 680 nm laser illumination on silicon detectors, the ionisation was only created in the first 10 μm of the diamond sample. For ^{241}Am the two alpha particle energies are 5485 keV (84,5 %) and 5443 keV (13 %) [69].

For the organisation of first measurements with diamond detectors, Heinz Pernegger was contacted. He made one of the first TCT measurements with an alpha source on high quality diamond at room temperature [70]. The discussions motivated to modify the RD39 setup [68] to allow for measurements with alpha source down to 57 K and first results on diamond were obtained.

For measurements with an alpha source in liquid helium, the range of the alpha particles in the liquid needed to be estimated. Superfluid helium at 1 atm and 1.8 K has a density of 147 kg/m^3 [25]. The total stopping power for a 5.5 MeV alpha particle in Helium is of 887 MeV·cm^2/g and their corresponding range is of $3.9 \cdot 10^{-3}$ g/cm^2 [71]. For the mentioned density in liquid helium, the range of the alpha particle is therefore of only 260 μm.

For first tests in liquid helium, the ^{241}Am alpha source was placed as close as possible ($\sim$900 μm) to the detector. The source and the detector were immersed in the liquid helium and after temperature equilibration, the setup was placed a few millimetres above the liquid surface. This allowed to see very first pulses at a temperature of 4.2 K.

Hendrik Jansen together with members of the cryogenic laboratory improved the setup [72]. The detector and the source were installed in a separate leak-tight container, with good thermal contact to the walls. In this container the pressure was adjusted in such a way that the alpha particle is not stopped and that the residual gas acted as contact gas to reach the wanted temperatures.

The charge collection decrease with temperature in a single crystal CVD diamond detector from an alpha particle can be seen in figure 4.18. Due to these measurements, it was feared that diamond detectors would not be suitable for the operation at liquid helium temperatures. In chapter 5, it will be shown that this pronounced decrease in collected charge is not the case for minimum ionizing particles (MIPs). The measurements with alpha particles are special, as the charges are generated within a very small region of the detector, whereas MIPs generate charges over the entire thickness of the detector. The charge density in the case of the alpha source can be estimated to be of 10^{21} cm^{-3}, leading to a dense plasma in which the outer charges screen the applied electric field from the inner charges. It was shown that the reduction in charge in the case of alpha particles is due to the generation of bound states between electrons and holes within the very dense charge cloud [72]. Such a bound state between an electron and a hole is referred to as an exciton. Its evaporation lifetime is temperature dependent and increases from ∼30 ps at 300 K to ∼150 μs at 50 K [72]. Charges bound in such a state are only measurable after the evaporation of the exciton, which hence explains the observed reduction in collected charge.

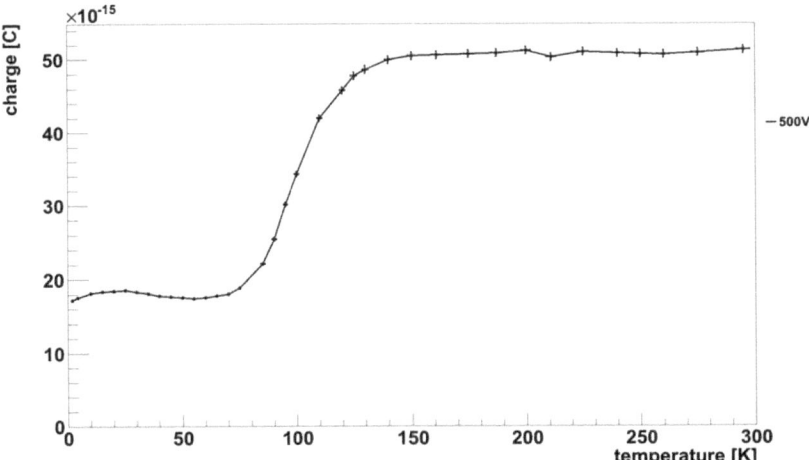

Figure 4.18: Collected charge induced by alpha particles from an ^{241}Am source in single crystal CVD diamond detector under 500 V for different temperatures. Courtesy Hendrik Jansen.

The results of the charge drifts through the bulk are summarised in table 4.1. While at room temperature the electron and hole low field mobilities are larger in diamond material than in silicon sensors, the situation is inverted at liquid helium temperatures. The saturation velocities at liquid helium temperatures are larger for both types of charge carriers in diamond material and the results from [73] are shown in figure 4.19.

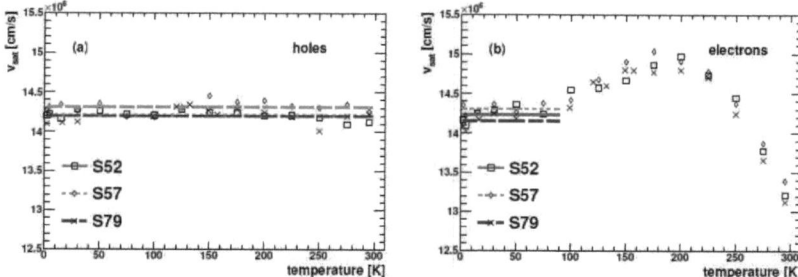

Figure 4.19: The diamond saturation velocity for three sCVD samples (S52, S57 and S79) is plotted against the temperature for holes (on the left) and electrons (on the right) [73]. The hole saturation velocity is constant over the measured temperature range, whereas the electron saturation velocity is not.

4.5 Charge carrier characteristics in liquid helium

When a particle traverses the liquid helium, it ionises the helium atoms. The ions and electrons then travel under the applied electric field inducing the signal. To understand the signal formation, it is necessary to consider the behaviour of the ionisation charges in liquid helium. The further discussion will show that this behaviour is peculiar.

The ion mobility μ can be estimated according to the Einstein-Smoluchowski equation [74]:

$$\mu = \frac{qD}{k_B T} \tag{4.9}$$

where T is the absolute temperature, q the charge of the ion and D the diffusion coefficient. First experiments on diffusion coefficients in liquid helium have lead to surprising results by showing that the drift of charge carriers in the liquid must underlie special principles. The experimental values for the diffusion coefficient of ^{3}He in ^{3}He-^{4}He mixtures are about two orders of magnitude higher than for positive ions at 1.2 K. This is surprising as the expectation would have been, that the ion and the ^{3}He atom are similar objects with respect to their mobilities within the liquid.

Theoretical explanations could be found to describe these surprising measurements. It was found that the positive ion is surrounded by a solid cluster of He atoms, often referred to as "snowball" [75], whereas the negative ion is an electron inside an expansion of the liquid, called "e-bubble" [76]. These structures will further be explained.

4.5.1 Positive Ions

In the Atkins model from 1959 the ion produces a strong non uniform electric field, which polarizes the neighbouring fluid. To minimize the energy, the polarised atoms are attracted by the ion, which causes an increase of the local fluid density, that can be estimated. Calculations of the expected pressure around the ion show that the helium is not fluid any more, but a solid phase builds up around the ion [75], therefore the name "snowball".

In his model, Atkins obtains a "snowball"-sphere with a radius of about 6 Å and an effective mass of 100 ^{4}He atoms [75]. Newer measurements and calculations find the ion-

	Si-10kΩcm-p$^+$-n-n$^+$	sCVD
RT e$^-$ mobility [cm^2/Vs]	1450 Section 4.3	1802 ± 14 [73]
RT h mobility [cm^2/Vs]	450 Section 4.3	2534 ± 20 [73]
LHe e$^-$ mobility [cm^2/Vs]	$(8.27\pm0.66)\cdot10^4$ Section 4.3	3061 ± 55 [73]
LHe h mobility [cm^2/Vs]	$(2.37\pm0.6)\cdot10^4$ Section 4.3	10 736 ± 244 [73]
RT e$^-$ saturation velocity [cm/s]	$(8.66\pm0.15)\cdot10^6$ Section 4.3	$(1.32\pm0.02)\cdot10^7$ [73]
RT h saturation velocity [cm/s]	$(7.18\pm0.09)\cdot10^6$ Section 4.3	$(1.42\pm0.04)\cdot10^7$ [73]
LHe e$^-$ saturation velocity [cm/s]	$(1.20\pm0.01)\cdot10^7$ Section 4.3, [54]	$(1.42\pm0.01)\cdot10^7$ [73]
LHe h saturation velocity [cm/s]	$(6.67\pm0.03)\cdot10^6$ Section 4.3	$(1.42\pm0.04)\cdot10^7$ [73]

Table 4.1: Charge carrier characteristics comparison between the sCVD and the 10 KΩcm p$^+$-n-n$^+$ silicon detector. RT denotes room temperature, LHe liquid helium temperatures, h holes and e$^-$ electrons.

snowball radius to be between 4 and 6.6 Å corresponding to 40 and 60 ^{4}He atoms [77, 78], depending on temperature and pressure. In the case of positive ions, the radius is found to be increasing with decreasing temperatures.

4.5.2 Negative Ions

Atkins expected his model to be true for positive and negative ions. His model of the ions in liquid helium is in agreement with experiments for positive ions as mentioned before, but it can not explain the difference seen between positive and negative ions. Therefore a "bubble"-model was suggested by Kuper in 1961 [76].

In Kuper's model the electron bubble is due to a balance between the quantum-mechanical zero-point (ground-state) energy of the electron, the surface energy of the bubble wall, the pressure-volume work proportional to the applied pressure, and the polarization energy of the helium atoms.

Kuper's calculations lead to a bubble radius of 12.1 Å and an effective mass of 100 ^{4}He atoms [76]. Recent estimations are that the radius of the "e-bubble" sphere is of 15 to 20 Å [79] with an effective mass of several hundreds of ^{4}He atoms [77]. The radius is again temperature and pressure dependent and increases with temperature, contrary to the positive ion.

A free electron injected through the surface into the liquid helium thermalizes within a few picoseconds into the charged bubble state.

4.5.3 Electron-ion recombination

For the operation of a liquid helium chamber as a beam loss monitor, the expected impact of the recombination between ions and electrons on the charge yield needs to be known.

The recombination process starts with the approach of the electron and the positive ion. The Coulomb attraction between the charges increases with r^{-2}. The hypothesis is that the actual recombination then happens by electron tunnelling out of the "e-bubble" through the walls of helium atoms into the "snowball" helium ion.

Experiments showed that no recombination fluorescence from states closer than 1.8 eV to the ionisation threshold exists [81]. This can be explained by the experimentally obtained value for the binding energy of the electron in the bubble, which is of about 1 eV [80], together with the relaxation of the bubble structure in the helium environment after recombination, which results in an additional energy loss of about 1 eV. These points are an indication that the recombination between the bound electron in the bubble with the ion requires an energy of about 1.8 eV.

To understand what impact electron-ion recombination might have in liquid helium, one can use the Langevin equation, which allow to estimate the recombination coefficient α through the electron (μ_-) and ion (μ_+) mobilities [82]:

$$\alpha = e(\mu_- + \mu_+)/\epsilon_0 \qquad (4.10)$$

With the charge mobilities from section 4.5.6, one can hence estimate the recombination coefficient to be of $3 \cdot 10^{-7}$ cm^3/s, corresponding to a recombination cross section of 10^{-9} cm^2 in liquid helium [80, 83], while for gas the value is of 10^{-15} cm^2. In the case of an alpha source in liquid helium, the initial density of generated charges is large. The characteristic time of recombination is estimated to be of only 0.1 ns in this case [82]. For protons with minimum ionising energies this situation should be different as the charge density is much lower and therefore significantly less immediate recombination should occur.

4.5.4 Breakdown field in liquid helium

An electric breakdown of the liquid helium chamber has to be avoided, as it possibly leads to heat dissipation, damage of instruments and unwanted dumps of the LHC beam.

Depending on the electrode material, its size and its shape different values for the breakdown voltage of liquid helium can be found in literature showing very large variations [84]. In figure 4.20 the breakdown fields are indicated on the helium phase diagram. The breakdown values obtained in these experiments range from 10 kV/cm to above 142 kV/cm.

For comparison the breakdown voltage in air is of 45 kV/cm (again depending on the electrode material, its size, its shape and the humidity of air).

A safety factor of 2 between the lowest reported breakdown field of 10 kV/cm at 1.8 K and the applied voltage during beam tests has been taken into account. During the operation of the LHC, the safety electric field should not be exceeded to avoid damage to the electronic equipment and false beam dumps.

In case of a magnet quench, the voltage on the chamber has to be switched off automatically, as the helium conditions between the plates might be unstable.

4.5.5 Estimation of the ionised charge per MIP

The electronic stopping power $S(E)$ of helium for protons with a momentum of 9 GeV/c is of 2.14 MeV·cm^2/g [86]. The density ρ of liquid helium is of 0.145 g/cm^3. The liquid

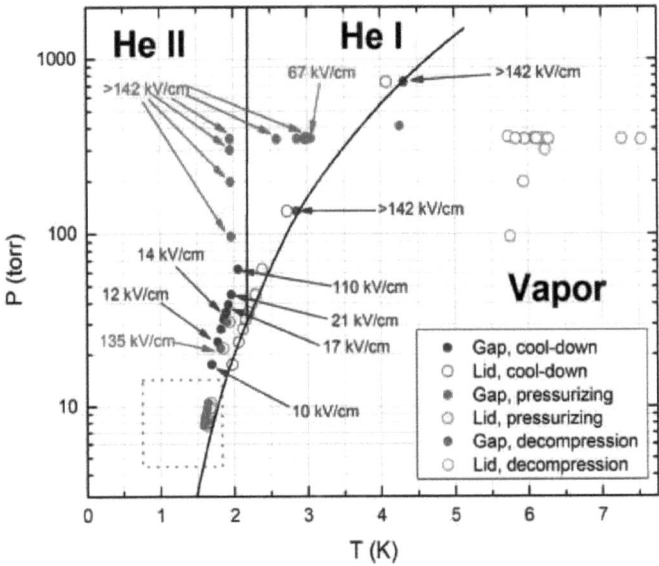

Figure 4.20: Breakdown voltages depicted inside the phase diagram of Helium [85].

density does not significantly change within the measurement ranges of temperatures from 1.5 K to 4.2 K and pressures between 1 mbar and 1.1 bar [25]. The W-factor gives the average energy required to create an electron-ion pair. In gaseous helium the W-factor is approximately 43.5 eV [87, 88] and strongly depends on the impurities in the gas (Jesse effect). As mentioned in section 3.3.2, the effect of impurities should not be an issue for the application due to demanding cleaning procedures for the helium liquefaction and due to the properties of superfluid helium. No value for the W-factor of liquid helium could be found in literature. Together these values allow an estimation of the expected charge from the proton beam. The number of ionised electrons n_e can be estimated using the following relation:

$$n_e = \frac{S(E)\rho}{W} \qquad (4.11)$$

A proton with a momentum of 9 GeV/c leads therefore to a generation of 7130 ionised electrons in one cm of liquid helium, assuming a $W = 43.5$ eV. This corresponds to a charge of 1.14 fC/cm. For the first liquid helium chamber prototype with an active length of 3.9 cm, this leads to a charge of 4.45 fC per proton.

4.5.6 Estimation of the charge collection time

The special structure of ions and electrons in liquid helium with an effective mass of 40 to several hundreds of helium atoms results in a low charge carrier mobility. This needs to be taken into account for the design of the detector and for its timing properties.

The electron mobility in superfluid helium at 1.8 K is of 8 mm^2/Vs (=0.08 cm^2/Vs), while for ions it is of 10 mm^2/Vs [77]. For comparison, the earlier found electron mobility in silicon detectors at liquid helium temperature is of $(8.27\pm0.66)\cdot10^7$ mm^2/Vs, while in other fluids like liquid argon it is of 47000 mm^2/Vs [89]. The ion mobility in liquid argon on the other hand is of only 0.28 mm^2/Vs [90].

From the charge mobilities μ and the applied electric field, one can calculate the time t for full charge collection of the LHe detector:

$$t = \frac{d}{\mu E} \quad (4.12)$$

Where d is the gap between the electrodes and E is the electric field.

The charge drift time for a chamber with 1 mm distance between the plates can hence be estimated, as shown in figure 4.21. This value overestimates the charge collection time, as for a MIP the charges are created evenly over the whole distance between the plates.

In order to guarantee a complete charge collection below 1 ms, a minimum electric field of 500 V/mm hence needs to be applied at 4.2 K, while for 1.8 K the value is of 125 V/mm.

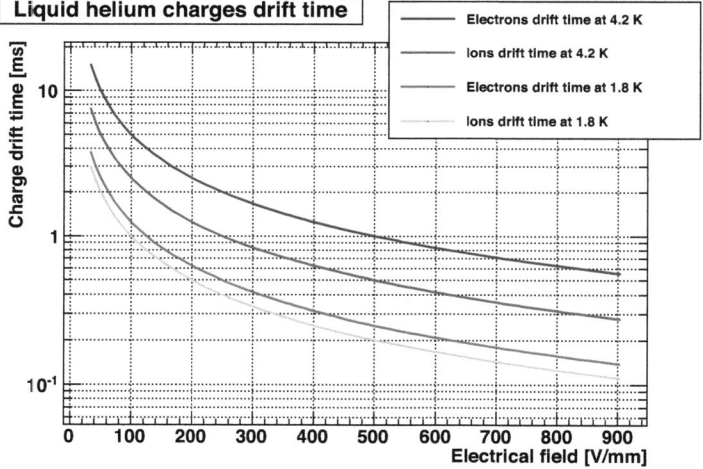

Figure 4.21: Estimated charge drift time in 1 mm of liquid helium.

4.5.7 Magnetic field considerations

The magnetic field at the detector position is estimated to be of 2 T. If the magnetic field is not parallel to the electric field and affects the movements of the charges, it therefore worsen the charge collection time. The movements of the charges can be described using the Lorentz force:

$$\vec{F} = q[\vec{E} + (\vec{v} \times \vec{B})] \quad (4.13)$$

A worst case scenario by maximising the effect of the magnetic field on the charges is further considered to estimate the consequences on the charge drift time. The largest charge mobility of interest is the one of ions in superfluid helium at 1.8 K. It is of 10 mm^2/Vs. With an electric field of 300 V/mm, the drift velocity of the charges will be of 3 m/s. In the worst case, the magnetic field lines are exactly perpendicular to the drift velocity of the charges. In this case the magnetic force on the charges due to the magnetic field can be estimated and is of 6 N/q. On the other hand the force from the electric field on the same charge is of 300 000 N/q. The effect of the magnetic field on the charge drift can therefore be neglected with respect to the timing properties of the liquid helium chamber compared to the one from the electric field.

4.6 Summary

Laboratory experiments were carried out to measure the charge carrier properties in silicon and diamond detectors down to liquid helium temperatures. It has been shown that the electron and hole drift mobilities and velocities are strongly temperature and voltage dependent. The electron and hole saturation velocity in silicon at 4.2 K is of $(1.20\pm0.01) \cdot 10^5$ m/s and $(6.67\pm0.03) \cdot 10^4$ m/s respectively. The low field electron and hole mobility at 4.2 K is of $(8.27\pm0.66) \cdot 10^4$ cm^{2}V^{-1}s^{-1} and $(2.37 \pm 0.6) \cdot 10^4$ cm^{2}V^{-1}s^{-1} respectively.

It has been further shown that the major downside of particle detection based on the liquid helium itself, is the slow drift of the charge carriers due to their formation of special structures in the liquid: "electron-bubbles" and "ion-snowballs". To collect all charges generated by a MIP in less than 1 ms, an electric field of at least 125 V/mm has to be applied at 1.8 K. The estimated charge per MIP in liquid helium is of 1.14 fC/cm.

CHAPTER 5

Cryogenic single particle detection with diamond and silicon detectors

A charged particle traversing the detector material generates pairs of charges. These charge carriers start to move under the influence of an applied electric field, which induces a measurable current on the detector contacts. In this Chapter the silicon and diamond detector properties with respect to single minimum ionising particle detection are presented. The measurements allow comparisons between the detectors and between the their operation at room temperature and at liquid helium temperatures.

5.1 Signal estimation from a MIP

The mean rate of energy loss $-\langle dE/dx \rangle$ by moderately relativistic charged heavy particles can be calculated through the Bethe-Bloch equation [91].

The Bethe-Bloch equation [91] is a function of the energy transfer and the excitation energy I:

$$-\left\langle \frac{dE}{dx} \right\rangle = K\rho q^2 \frac{Z}{A} \frac{1}{\beta^2} \left[\frac{1}{2} \ln\left(\frac{2m_e c^2 \beta^2 \gamma^2 T_{max}}{I^2} \right) - \beta^2 - \frac{\delta(\beta\gamma)}{2} \right] \quad (5.1)$$

where T_{max} is the maximum kinetic energy that can be transmitted to a free electron in a single collision. It is given as:

$$T_{max} = \frac{2m_e c^2 \beta^2 \gamma^2}{1 + 2\gamma m_e/M + (m_e/M)^2} \quad (5.2)$$

K/A is of 0.307 MeVcm^2/g for the atomic mass $A = 1$ g/mol, ρ is the density of the material, q is the charge of the incident particle, Z is the atomic number, β and γ are the reduced velocity v/c and the Lorentz-factor respectively, $\delta(\beta\gamma)$ is a density effect correction to ionisation energy loss, M is the incident particle mass and $m_e c^2$ is the electron mass in units of energy. The density ρ of silicon material is of 2.33 g cm^{-3}, while the one of diamond material is of 3.52 g cm^{-3}.

The mean charge per minimum ionising particle $<Q(0)>$ for a non-irradiated detector with thickness d can be calculated using the following equation:

$$<Q(0)> = \frac{-\left\langle \frac{dE}{dx} \right\rangle d}{w} \quad (5.3)$$

Detector	d [μm]	$< -dE/dx > /\rho$ [MeVcm^2/g]	w [eV]	$< Q(0) >$ [fC]
Si	300	1.67	3.6	4.3
sCVD	500	1.76	13	3.1

Table 5.1: Mean energy loss and mean generated charge in silicon and diamond detectors for a Minimum Ionising Particle (MIP) with $\beta\gamma = 3$. Note that the mean energy loss given in literature is normalised by the density.

where the energy w for the creation of an electron-hole pair is of 3.6 eV in silicon diodes and of 13 eV in diamond sensors.

The obtained values for $< Q(0) >$ are shown in table 5.1. Further information and a very detailed description of the used terms and formulas can be found in [44, 91].

5.2 Fluctuations of energy loss - Landau distribution

The energy loss probability distribution of particles with same momentum in detector materials can be described by a Landau distribution. This distribution is due to fluctuations of the amount of energy lost by the particles. The number of collisions of the incident particle with the detector material varies, as well each energy exchanged during the collisions. Going from thin detectors to thick absorbers the energy loss probability distribution turns from a Landau distribution to a Gaussian form.

The characteristic values of a Landau distribution are:

- the Most Probable Value (MPV), corresponding to the energy loss of highest probability,
- the σ, being a measure for the width of the distribution and
- the mean value of the distribution.

5.2.1 Effects of cuts on the Landau distribution

In the case of a counting mode read-out of the semiconductor detectors, a threshold needs to be set to distinguish between electric noise or background and an actual event from particle losses. Figure 5.1 shows two Landau distributions. The red distribution has a larger MPV than the blue one. Figure 5.2 shows how the threshold level influences the resulting apparent MPV and the apparent mean. The critical result of this analysis is the crossing point of the mean value. At a certain threshold level the distribution with smaller MPV has a larger resulting mean. This is critical when comparing two different detectors or the same detector at different temperatures. The situation is shown here in case of an implemented threshold level on the detector signals, but the results in case of external triggering have to be treated carefully too, because of a possible influence from the electric noise.

The measured mean of the distribution is also affected by the upper limit of the distribution. When measuring with an oscilloscope the amplitude divisions are set in a way to give best resolution around the mean of the distribution. Signals from extremely high ionisation events will hence be cut at the higher limit of the oscilloscope range.

In the further analysis and experiments, these issues were intended to be solved by using further detectors as external triggers. The representation of the MIP signal through the mean pulse of all pulses (as shown in the case of the laser measurements) is done for visualisation, but interpretations or conclusions should be done carefully. It is critical to keep this in mind

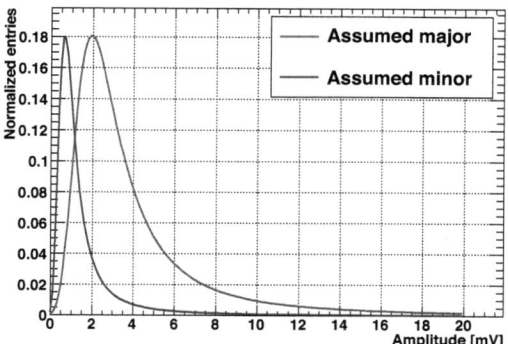

Figure 5.1: Two Landau distributions are assumed. The main difference between the two is the larger MPV of the red distribution.

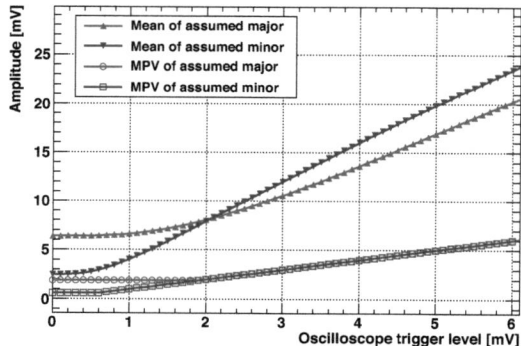

Figure 5.2: Evolution of the apparent mean and MPV of the two Landau distributions as a function of an applied threshold cut.

when analysing and comparing mean pulses from silicon and diamond detectors at different voltages and at different temperatures.

5.3 Cryogenic amplifier

5.3.1 Introduction

By installing the amplifier into the cryostat, the noise level can be significantly reduced and in addition the cable length from the detector to the amplifier can be avoided. Johnson-Nyquist noise is electronic noise generated by thermal movements of charges. Even without any current in the circuit, the random thermal motions of the electrons produce voltage fluctuations along the conductor. These fluctuations account for a mean square noise voltage at its open terminals. The noise voltage has a Gaussian probability density function with a mean value of zero. Johnson analysed the noise experimentally, while Nyquist did the theoretical investigation of this thermal noise. The relation for the mean square noise voltage $\bar{u^2}$ per hertz of bandwidth is the following [92]:

$$\bar{u^2} = 4k_B TR \qquad (5.4)$$

where T is the absolute temperature, R is the resistance and k_B is the Boltzmann constant equal to $1.38 \cdot 10^{-23}$ J/K. Per hertz of bandwidth the thermal mean noise voltage is therefore expected to be a factor 12.6 smaller at 1.9 K compared to room temperature.

A 40 dB current amplifier (C2, device number 88) from CIVIDEC was therefore tested inside liquid nitrogen (77 K) and liquid helium (4.2 K), to check the feasibility of using the device for future beam tests. The amplifier worked during 30 hours in liquid helium. It also proved its stability over 8 hours in a vacuum of about 0.1 bar at room temperature.

Therefore several tests were performed with the goal to conclude on the feasibility of placing the amplifier into the cold and identify its performance at these low temperatures. Further the long term stability of the device in cold and in vacuum was monitored.

5.3.2 Testing procedure

The components, the structure and the used material in the amplifier were expected to mechanically withstand the deformations in the cold. The first immersion of the amplifier into liquid nitrogen confirmed the mechanical stability of the amplifier.

Later a 300 ps pulser and and the PM 5786 pulser from Philips (pulse generator with 1 Hz-125 MHz and risetime 2 ns-0.1 s) were used to measure the gain at liquid helium temperatures. Figure 5.3 shows the tested 40 dB current amplifier on the right with a capacitance in the middle and the 300 ps pulser on the left.

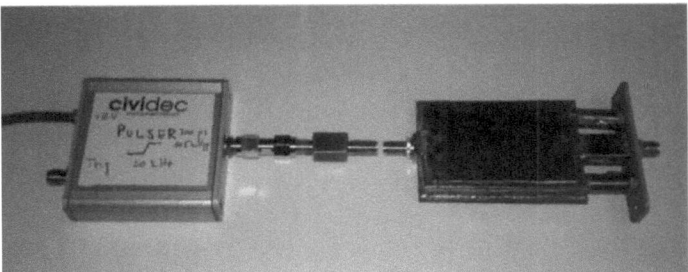

Figure 5.3: Tested 40 dB current amplifier on the right with a capacitance in the middle and the 300 ps pulser on the left.

Finally the amplifier was connected to a single crystal CVD diamond detector with a beta source and an alpha source and the signals were read out on the oscilloscope at different temperatures.

Picture 5.4 shows the detector holder and the amplifier. An alpha source was placed on the holder and the whole setup was lowered into the cryostat, which was later filled with liquid helium. The Cernox temperature sensor was placed on the amplifier, as it is the instrument of interest.

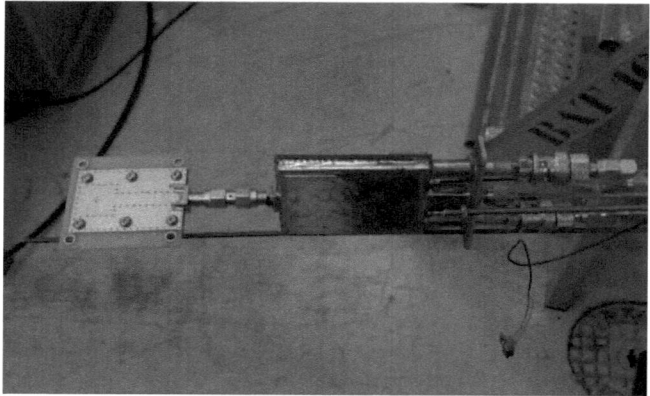

Figure 5.4: The diamond detector holder from CIVIDEC and the amplifier prepared for the tests in liquid helium.

Bandwidth	Stdev [mV]
200 MHz	0.28
3 GHz	0.57

Table 5.2: Electronic noise characteristics of the oscilloscope without connection.

5.3.3 Electronic noise variation

First the electronic noise of the used 3 GHz LeCroy Oscilloscope without any connection was measured. The input impedance of the oscilloscope is set to 50 Ω DC for all measurements. The results are shown in table 5.2.

Inside the liquid helium cryostat, the electronic noise level was measured with the detector setup and the results are summarised in table 5.3. These noise levels are close to the values of the oscilloscope alone and therefore the oscilloscope becomes the main limitation in further noise reduction.

5.3.4 Gain variation

The characteristics concerning the gain were measured using the mentioned pulsers. These pulsers enabled a well defined signal at the input of the amplifier, which was compared on the

	Stdev [mV]	
Bandwidth	Room temperature	4.2 K
200 MHz	1.03	0.38
3 GHz	2.25	0.76

Table 5.3: Standard deviation of the electronic noise of the amplifier and the detector inside the cryostat.

Temperature [K]	296	77	4.2
Signal amplitude [mV]	64	45.6	22

Table 5.4: Signal amplitude at different temperatures.

oscilloscope with the corresponding output at different temperatures. During the cool down of the amplifier, a decrease of the gain was observed and the results of the amplitude variations are summarised in table 5.4. The gain reduction of the amplifier from room temperature to 4.2 K is of 66 %.

The measurement procedure was reproduced with a pulser from Philips, with the goal to generate pulses that better approximate the pulse shape expected from a minimum ionising particle, with respect to the timing properties. Figure 5.5 shows the input signal (in yellow), then the output at room temperature in dark yellow, the output at 4.2 K in red and between both the actual output at 110 K in blue and light yellow. Again a reduction of the gain from room temperature to 4.2 K of the amplifier of 66 % was measured. The shape of the signal further gives information that the bandwidth of the amplifier was reduced in the cold. This is visible in the rise and in the top part of the pulse. Also the undershoot was more pronounced in cold.

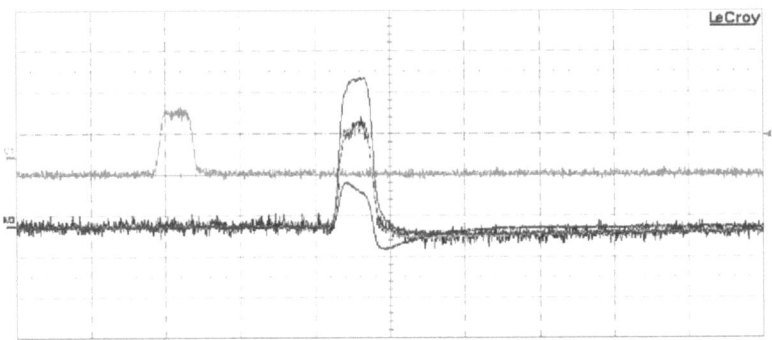

Figure 5.5: Amplifier output at different temperatures from an input signal (in yellow). The output at room temperature is shown in dark yellow, the output at 4.2 K in red and between both the actual output at 110 K in blue for the current shape and light yellow for the average curve.

This reduction in gain is equal to the reduction in electric noise from RT to 4.2 K. The expected advantage of installing the amplifier inside the cryostat is hence neutralised.

The amplifier gain can be increased by changing the supply voltage. At a constant temperature of 4.2 K, the supply voltage on the amplifier was hence increased with the goal to augment the amplifier gain and therefore reach a level comparable to the one at RT.

The signal shape for a supply voltage of 15 V is depicted in figure 5.6. The gain can be augmented, but the pulse shape distortion becomes even more pronounced. Such a distortion is not acceptable anymore for the use in a beam test experiment, as the physical results would be modified and even dominated by the used electronics. In addition the power dissipation was measured to be of 1.2 W at 15 V, which is too large for the use in liquid helium experiments.

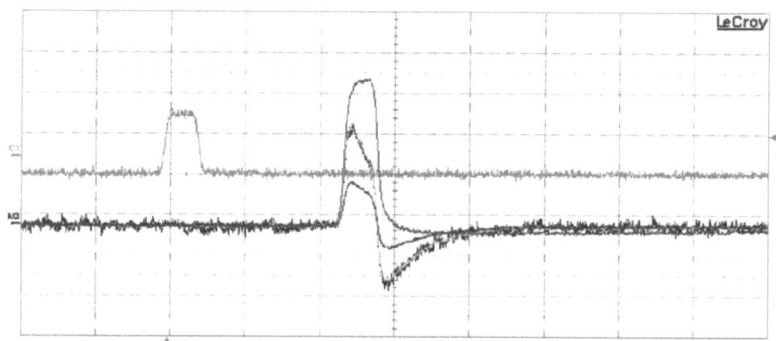

Figure 5.6: Amplifier output from an input signal (in yellow). The output at room temperature at 12 V is shown under memory 1 (in dark yellow), the output at 4.2 K and 12 V under memory 2 (in red) and between both the actual output at 4.2 K and 15 V supply voltage (in blue for the current shape and light yellow for the average curve).

Temperature [K]	Current [A]	Power [W]
295	0.09	1.1
130	0.07	0.8
100	0.07	0.8
77	0.07	0.8
40	0.06	0.7
20	0.06	0.7
10	0.05	0.6
4.2	0.04	0.5

Table 5.5: Power consumption of the amplifier at different temperatures.

5.3.5 Power consumption

With lower temperatures the power consumption (and therefore the heat dissipation) of the amplifier decreased, while the supply voltage was kept constant at 12 V. During the cool down by liquid helium inside the cryostat, the values from the voltage supply were recorded and can be found in table 5.5.

5.3.6 Further measurements

With β source at 77 K

During the amplifier tests with liquid nitrogen, measurements with a single crystal diamond and a β source (^{90}Sr source with a maximum energy of 546 keV) have been performed. Example pulses proving the functionality of the entire setup can be seen in figure 5.7. This was important to check, as beam tests with minimum ionising particles are performed with the same components.

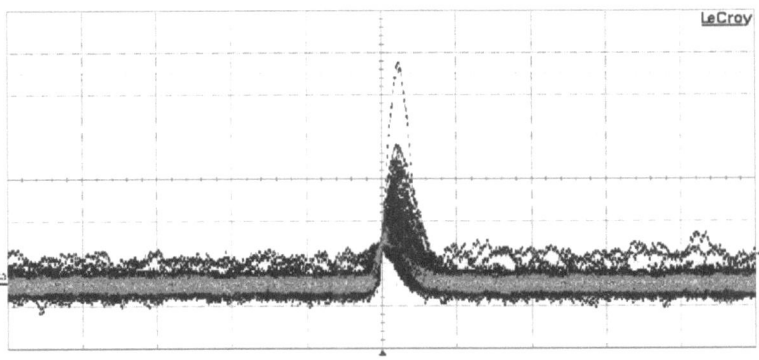

Figure 5.7: Pulses from a β source on a single crystal diamond detector and with the amplifier immersed in liquid nitrogen at 77 K.

With α source at 4.2 K

Due to radiation protection, only an α source is available for temperatures down to liquid helium.

The signals from alpha particles in sCVD at 4.2 K are shown in figure 5.8. Comparing the pulse widths from electron and hole drifts, one can deduce the higher drift velocity of holes compared to electrons in the diamond material at 4.2 K, as measured and illustrated in detail in [72].

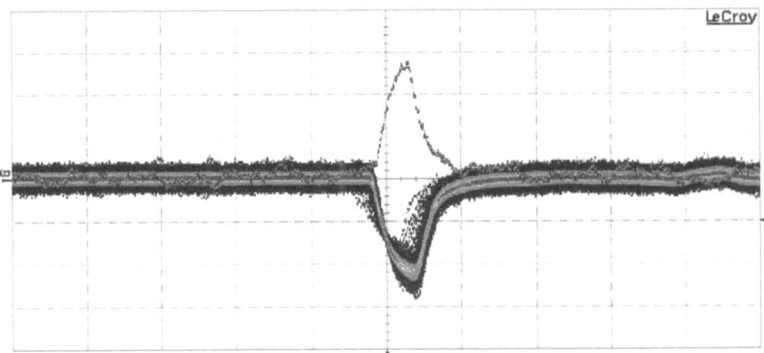

Figure 5.8: Alpha particle signals measured with a sCVD and the amplifier installed inside the cryostat at 4.2 K. The positive and negative pulses correspond to the holes and electrons drift respectively traversing the diamond detector bulk. First a few exemplary hole pulses were recorded and then the voltage was switched to measure the pulses from the electron drift.

5.3.7 Cryogenic amplifier summary

The used 40 dB current amplifier from CIVIDEC withstands temperatures down to liquid helium at 4.2 K and works without any noticeable difference in a vacuum of 0.1 mbar.

The noise level of the electronics could be largely reduced at 4.2 K. Unfortunately the noise reduction goes along with a reduction of the gain of the amplifier, which neutralises the awaited advantages. In addition to the reduction in gain, a distortion of the pulse shapes is observed, which is mostly due to a reduction of the amplifier bandwidth.

For a future development of a cryogenic amplifier the used components should be chosen in a way that the amplifier has a 40 dB amplification and a 2 GHz bandwidth at liquid helium temperatures. Important for the application in cold is further the power dissipation, which would need to be optimised in the case of this amplifier, as too much heat is produced for a cryogenic application.

Due to the reduction in gain and in bandwidth and the need of further feedthroughs on the cryostat, the final decision was to not install the amplifier inside the cryostat for the beam tests.

5.4 Single particle detection

5.4.1 Experimental setup

Single minimum ionising particle (MIP) measurements were performed with the semiconductor detectors, using 40 dB current amplifiers from CIVIDEC. The pulses were recorded on a 3 GHz LeCroy oscilloscope with 200 MHz bandwidth limitation. This bandwidth limitation was chosen to reduce the electric noise level (as shown in table 5.2), without affecting the timing information of the signals.

5.4.2 Beam line

Tests with protons were performed in the PS East Area, using the beam from the PS. The conditions in the T9 beam line were:

- Particle momentum of 9 GeV/c,
- Beam spread with a symmetric FWHM of 1 cm at the cryostat,
- Maximum beam intensity per spill of $3.5 \cdot 10^5$ protons/cm^2, adjustable to lower intensities with the use of horizontal and vertical collimators.

Room temperature setup

The room temperature setup for single particle detection is shown in figure 5.9. The measurements were done on silicon detectors, single crystal CVD diamond detectors (sCVD), polycrystalline CVD diamond detectors (pCVD) and Diamond On Iridium (DOI) [93] detectors.

Examples of these measurements with a charge sensitive amplifier are shown in figure 5.10.

Cryogenic setup

The cryogenic setup was less optimal for precise measurements of single particles compared to the room temperature setup because of the 1.5 m cables between detectors and amplifiers.

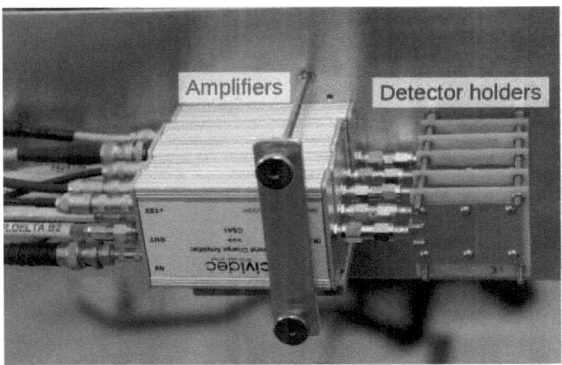

Figure 5.9: Setup for precise single particle detection measurements at room temperature. The outer two detectors are the smallest in active area and are used as external triggers for the detectors in the middle, whose signals are recorded and analysed.

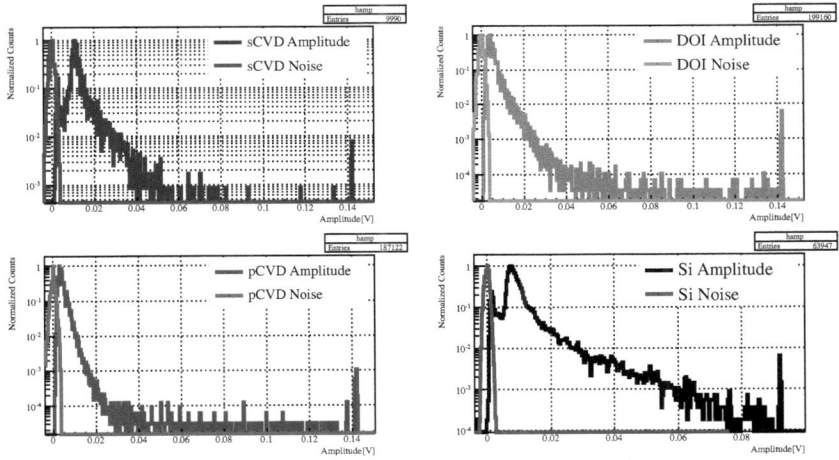

Figure 5.10: Single particle distributions at room temperature measured with sCVD, diamond on iridium (DOI), pCVD and Si (from left to right) using a CSA from CIVIDEC (C6) - Courtesy of Christina Weiss. On the x-axis is the amplitude in [mV] and on the y-axis are the normalised counts. For Si and sCVD the MPV of the distribution is above the noise.

Several amplifiers were tested for the measurements under cryogenic conditions. With all tested Charge Sensitive Amplifiers (CSA), the reflections due to the impedance mismatch and the cable length were dominant and could therefore not be used for reliable measurements. With the 40 dB current amplifier from CIVIDEC (C2) the SNR is worse compared to the CSA, but the reflections have a minimal impact on the signal. The final setup for the beam tests can be seen in figure 5.11.

Figure 5.12 shows the installation of the amplifiers on the top of he cryostat. At the be-

Figure 5.11: Detector installation for the low temperature measurements.

ginning, room temperature tests were performed for reference measurements and calibrations. Afterwards the cryostat was filled with liquid helium to measure the detector properties at 4.2 K and below.

Figure 5.12: Amplifier installation at the top of the cryostat. The left picture shows the adjustment for room temperature calibration measurements at the beginning of the beam test. The right picture shows the situation one week later when measuring in cold.

Figure 5.13 shows the experimental area with all equipment for the detector tests in the beam. For the measurements in liquid helium, the vacuum pump was turned on and left running for about 4 hours. The pump generated vibrations, affecting the noise level of all detectors. For the presented measurements and results below 4.2 K, the pump was therefore turned off and the measurements could be performed as long as the temperature stayed stable. As soon as the temperature started increasing, the measurements were stopped and the pump was turned back on.

All relevant cryogenic information, like temperature (measured with Cernox), pressure and liquid helium level, is recorded.

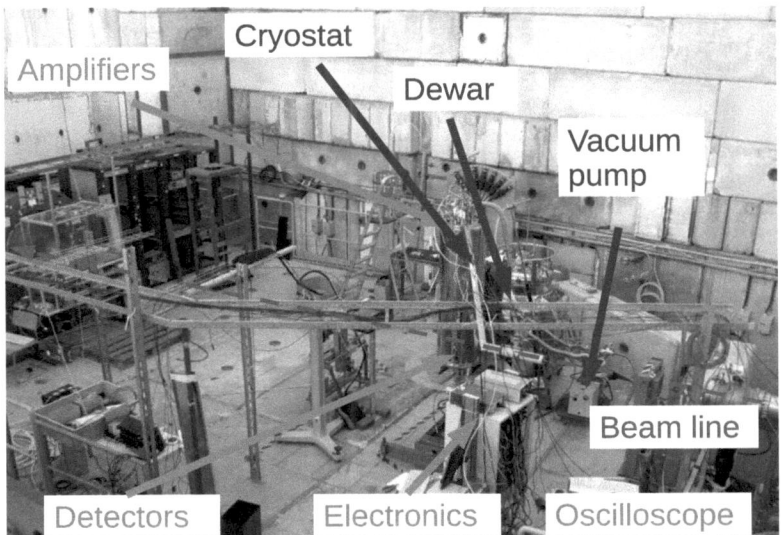

Figure 5.13: Overview over the experimental area with the installation of the cryogenic systems and the electronic read-out. The cryostat is fixed on a moveable table.

Detector	Thickness [μm]	Active area [mm^2]	Metallisation
Si-10kOhm-T9	300	7x7	Al
sCVD-T9-1	500	4.7x4.7	Ti + Au
sCVD-T9-2	500	4.7x4.7	Ti + Au
sCVD-T9-3	500	4.7x4.7	Ti + Au
pCVD-T9	500	8x8	Ti + Au

Table 5.6: Detector samples for the measurements in the low intensity beam.

Detector samples

For the beam test a p^+-n-n^+ silicon detector Al metallisation was used. The thickness of the sample was of 300 μm and its active area of 7x7 mm^2.

In addition 3 single crystal CVD diamond detectors (sCVD) and one polycrystalline CVD diamond detector (pCVD) with a double layer metallisation of titanium and gold and a thickness of 500 μm were used. The active area of the sCVDs was of 4.7x4.7 mm^2 and the one of the pCVD is of 8x8 mm^2.

The tested detector samples are summarised in table 5.6.

5.4.3 Results

In this section, the measurements of the tested semiconductor detectors with respect to single particle detection in liquid helium temperatures are shown. The relevant parameter of interest for the CryoBLM project is the mean value of the energy loss of a particle traversing the detector bulk. To obtain the basic detector properties and precisely compare the detectors at different temperatures with each other, the Most Probable Value (MPV) of the signal amplitude distributions needs to be considered. It was not possible to experimentally obtain the MPV above the electric noise at liquid helium temperatures. Still the measurements allow to draw many application relevant conclusions that will further be discussed. Additionally the particle showers from losses in the LHC lead to the detection of several particles in the CryoBLM detectors, in which case the determination of the mean value is less affected by the noise and the background, compared to single particle detection.

Observations with external trigger

The resulting distributions for MIP signal amplitudes with external trigger and maximum noise amplitudes are shown in figures 5.14, 5.15 and 5.16 for silicon detectors, sCVD and pCVD respectively. The distributions are made from a total number of 34000 to 35000 counts. In each case, the distributions are shown for room temperature and liquid helium temperatures. These results can not be directly compared to the ones in figure 5.10, because different amplifiers were used as explained in section 5.4.2. The MIP signal amplitude was the maximum current from the tested detector traversed by a particle. The presence of a particle going through the detector was signalised by an external trigger. The noise measurement was recorded in a time window without particles and the maximum value was taken out of this period. This time window was equal to the 40 ns time range in which the maximum with particle was obtained. The total number of counts from the noise distribution was equal to the one of the MIP signal amplitude distribution (meaning between 34000 and 35000 counts).

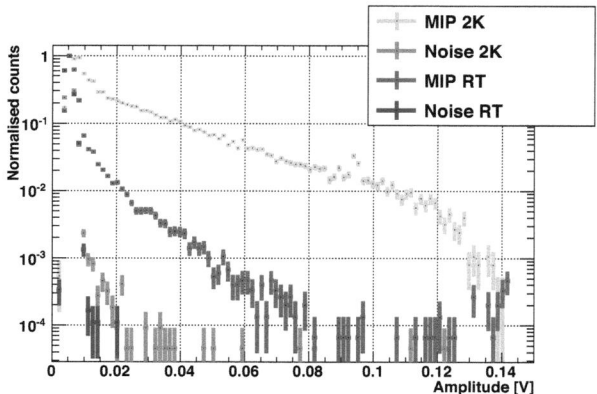

Figure 5.14: MIP signal amplitudes and maximum noise amplitudes for the silicon detector at RT and 2 K with 100 V applied. The mean value of the distribution is of 7.7 mV at RT and of 24.4 mV at 2 K.

The obtained MPV of the MIP signal distributions remained below the electronic noise for all cases, corresponding to a SNR below 1. The signal amplitudes increased for all detectors when going from 295 K to 2 K, which was due to the increased drift velocity of the charge carriers in the cold, as summarised in Chapter 4. The obtained mean signal amplitudes are summarised in table 5.7. This current amplitude increase was especially pronounced in the

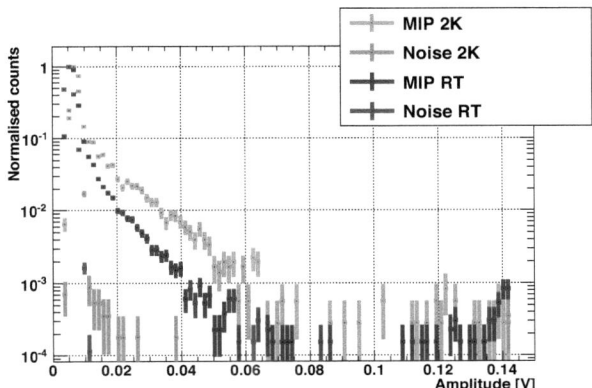

Figure 5.15: MIP signal amplitudes and maximum noise amplitudes for sCVD at RT and 2 K with 400 V applied. The mean value of the distribution is of 7.6 mV at RT and of 10.8 mV at 2 K.

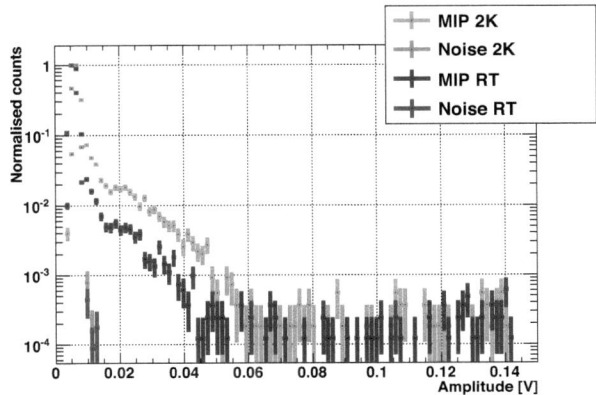

Figure 5.16: Signal amplitudes and maximum noise amplitudes for pCVD at RT and 2 K with 400 V applied. The mean value of the distribution is of 6.9 mV at RT and of 9.9 mV at 2 K.

case of the silicon detector, which makes silicon diodes the more efficient material for fast single particle detection at liquid helium temperatures, compared to sCVD and pCVD.

The integration of the MIP signals leads to the total charge generated in the detector by the MIP. This area is difficult to obtain and analyse due to the amplifier noise together with its AC based readout. The exact pulse width can not be determined if the signal amplitude lies within the noise. The pulse width determination is only possible for pulses above a certain threshold, in which case a cut would be performed on the Landau distribution and hence the obtained results would be biased.

The SNR is planed to be improved with the development of optimised charge sensitive

Detector	Signal amplitude at 295 K [mV]	Signal amplitude at 2 K [mV]
Si-10kOhm-T9	7.7	24.4
sCVD-T9-1	7.6	10.8
pCVD-T9	6.9	9.9

Table 5.7: MIP mean signal amplitudes for the tested detectors at RT and at 2 K.

amplifiers for measurements with this cable length and through the development of suitable cryogenic amplifiers. Both developments are currently under discussion.

Observations with signal threshold on measured detector

This section is complementary to the first section 5.2 of this chapter, where the threshold influence was analysed theoretically and assumption based. More practical and application relevant data is presented here, where a threshold was set on the measured detector, as it would be the case in a counting mode operation.

Figures 5.17 and 5.18 show the average MIP signal in a silicon detector and a single crystal CVD diamond detector when a threshold is applied.

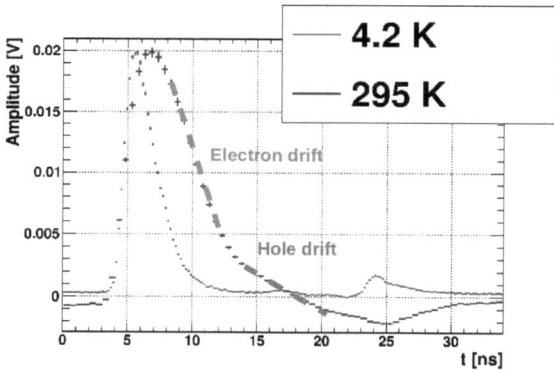

Figure 5.17: Silicon detector average pulse shape from MIP particles measured with a threshold of 6 mV at RT and at cryogenic temperatures. For the silicon detector signal at RT the drift of the slower holes can be distinguished from the electrons. This was already observed during laboratory measurements with the 1060 nm laser light, shown in figure 4.17. Reflections, due to slight impedance mismatch between detector and amplifier input, appear with 20 ns delay.

The silicon detector's Full Width Half Maximum (FWHM) of the signal from a MIP is of only 2.5 ± 0.7 ns and for the diamond detector the FWHM is of 3.6 ± 0.8 ns at liquid helium temperatures. At RT, the FWHM of the detectors is of 5.6 ± 1.2 ns and 5.1 ± 0.9 ns for the silicon and the diamond sensor respectively. The interpretation of the amplitude of the mean signals has to be done carefully, as has been shown in figure 5.2. The application of a threshold on the Landau distribution has a significant impact on the measured apparent mean value.

Figure 5.19 shows the silicon detector signals at room temperature and below for a not fully depleted detector on the left and for a fully depleted detector on the right. At 5 V only

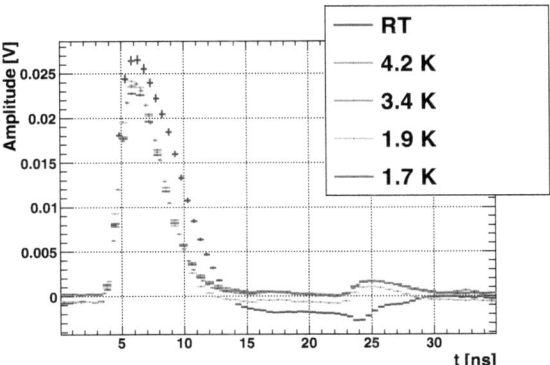

Figure 5.18: Single crystal CVD diamond detector average pulse shape from MIP particles measured with a threshold of 6 mV at RT and at cryogenic temperatures. Reflections appear again with 20 ns delay.

the MIPs with amplitudes from the high energy tail of the Landau distribution were detected. The observed signal shapes can be understood remembering the high mobility and high drift velocity of the charge carriers at liquid helium temperatures from Chapter 4. The detection efficiency of the detector at 5 V is of 15 % compared to higher voltages, which is shown in the right plot of figure 5.21. The left plot of the figure shows the variation of the normalised detector detection efficiency depending on the threshold setting.

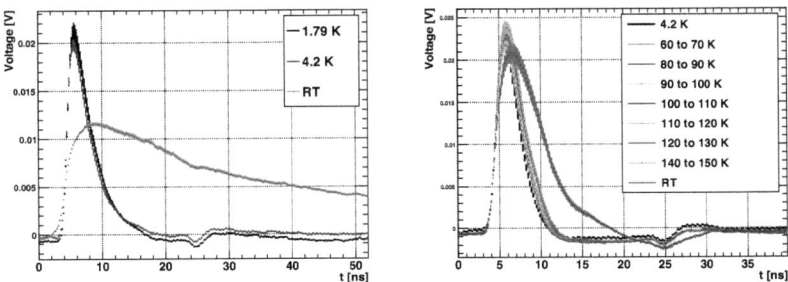

Figure 5.19: Silicon detector's average MIP signal at a 4 mV threshold and an applied voltage of only 5 V in the left plot and of 100 V in the right plot. The 5 V pulses are from the high energy tail of the Landau distribution. At 100 V the detector is fully depleted for all temperatures. Below 150 K the signal shape changes only slightly due to the increased drift velocity of the charges with decreasing temperature.

Figure 5.20 shows the MIP signals in a silicon detector at 4.2 K for different voltages. The major difference between the signals is the longer tail of the low field pulses and the low detection efficiency of the silicon detector at low fields. This similarity of the pulses is again due to the threshold and the high charge carrier mobility at liquid helium temperatures.

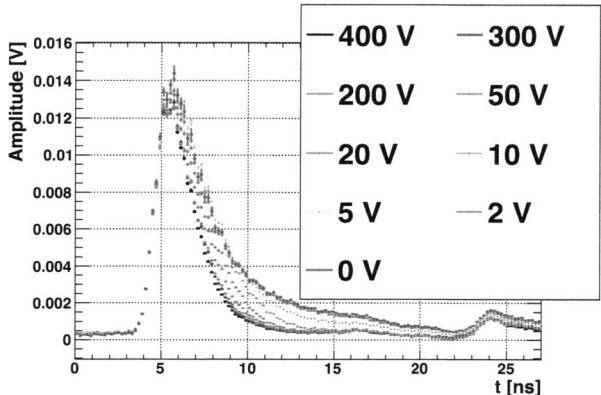

Figure 5.20: Average silicon detector pulse at 4.2 K and a threshold of 4 mV for different voltages. Due to the high charge mobility at low fields, the signal shape seems well developed at low applied voltages, but the detection efficiency is very low, as it is further shown in figure 5.21.

Due to the higher detection efficiency and the shorter signal pulses, silicon detectors have an advantage over diamond detectors, should the future CryoBLMs be operated in counting mode.

5.5 Summary

The measurements in the low intensity beam allowed the signal amplitude distributions at RT and at liquid helium temperatures to be obtained. An increase of the signal amplitudes with decreasing temperatures could be shown for all detectors and was especially pronounced for the silicon detector. An advantage of these increased signal amplitudes was that the detection efficiency of the detectors increased in this measurement setup.

It could further be measured that the silicon detector's FWHM of the signal from a MIP is of 2.5 ± 0.7 ns and that for the diamond detector the FWHM is of 3.6 ± 0.8 ns at liquid helium temperatures.

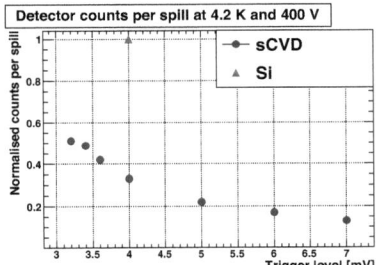 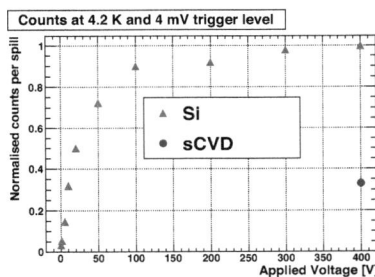

Figure 5.21: Normalised counts for different thresholds in the left plot and for different voltages in the right plot. The silicon detector has a three times better detection efficiency at 400 V, 4.2 K, 4 mV threshold and with this readout system. The relative detection efficiency depends on the threshold and the applied voltage.

CHAPTER 6

Radiation hardness of diamond and silicon detectors

This Chapter describes the radiation hardness of solid-state detectors and the experiments performed to investigate the detectors signal degradation at room temperature and at liquid helium temperatures. It was possible to derive a model for the reduction in the detector's charge collection due to radiation damage. This model is used to describe the measured data.

6.1 Radiation hardness

Solid state detector devices are damaged by particle irradiation. The radiation damage can be split into two main mechanism: displacement damage and ionisation damage. Displacement damage corresponds to modification of the crystal lattice due to irradiation, while ionisation damage corresponds to trapped charges and therefore parasitic fields within the detector.

The particles traversing the detector crystal lattice interact with the atoms and atom displacement is possible. This may lead to point defects in the crystal lattice like vacancies, foreign interstitial atoms, self interstitial atoms and foreign substitutional atoms (see figure 6.1). A Frenkel pair is one vacancy and one interstitial.

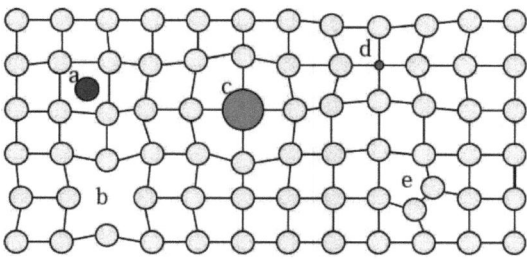

Figure 6.1: Schematic of possible point defects in the crystal lattice: (a) foreign substitutional atom, (b) vacancy, (c) and (d) foreign interstitial atoms and (e) self interstitial atom [44].

The charge carrier properties are strongly dependent on the point defect density in the bulk, as crystal defects correspond to a deviation of the periodicity of the crystal lattice and therefore influences the charge carrier mobility through scattering, its lifetime and its trapping probability.

In the Non-Ionising Energy Loss (NIEL) hypothesis the detector signal loss is proportional to the total energy absorbed, depending on the particle type and energy. The NIEL hypothesis is well understood and tested for silicon detectors [33] and the NIEL damage cross section is strongly related to the creation of previously described point defects. At low beam energies E the NIEL cross section is dominated by long-range Rutherford scattering. The cross section falls like $1/E^2$ and creates many small scale lattice displacements [94]. At intermediate energies (above few MeV) the anomalous elastic Rutherford scattering between the incoming particle and the atom nuclei in the sensor starts to be significant. The inelastic cross section dominates at energies above few hundreds of MeV. These collisions fragment the nuclei and the slow moving nuclear parts lead to strong lattice defects. Impurities in the silicon detector may lead to significant deviations from the NIEL scaling hypothesis. The group around M. Guthoff and the RD42 collaboration is currently working on the development of a NIEL model for diamond detectors.

The average energy to create a lattice displacement is of 25 eV in Si [56] and between 37-47 eV for diamond material [57]. At room temperature more than 90 % of primary created defects may recombine already during irradiation. In silicon material, post-irradiation defect migration takes place at room temperature, while damaged diamond sensors are more stable with respect to defect evolution with time, due to the high activation energies for migration of interstitials and vacancies [43].

The consequences of the described radiation damages on the operative silicon detector characteristics are changes in:

- the full depletion voltage,
- the leakage current and
- the CCE as a function of the applied field.

These changes lead to an unwanted reduction of the signal to noise ratio.

For diamond detectors the situation is similar with the exception that the leakage current does not increase, which is one of the main advantages of diamond detectors at room temperature.

6.1.1 Derivation of the charge collection degradation model

The reduction of the detector's sensitivity with increased radiation damage, corresponding to a reduced charge carrier life-time, can be modelled. The model is described in this section and the final equation 6.6 will be used later to fit the charge collection degradation curves obtained during the irradiation measurements.

Charges $Q(x)$ starting to move at a distance 0 from the detector contact, moving to a distance ξ (similar to the TCT method described in section 4.1) within the detector bulk of thickness d, induce a charge $Q_{ind}(\xi)$ on the contact:

$$Q_{ind}(\xi) = \frac{1}{d} \int_0^\xi Q(x) dx \qquad (6.1)$$

Where x is the coordinate between the detector contacts. $Q(x)$ can be rewritten as a function of the initial charge Q(0), time $t(x)$ and charge carrier lifetime τ, meaning that the initial charge decreases exponentially due to the trapping of charge carriers in crystal defects:

$$Q(x) = Q(0)\exp(-\frac{t(x)}{\tau}) = Q(0)\exp(-\frac{x}{\mu E \tau}) \quad (6.2)$$

The second transformation was obtained using the relation $t(x) = \frac{x}{\mu E}$, where μ is the charge carrier mobility and E is the applied electric field. Integration of the equation 6.1 leads to:

$$Q_{ind}(\xi) = Q(0)\frac{\mu E \tau}{d}(1 - \exp(\frac{\xi}{\mu E \tau})) \quad (6.3)$$

This last equation is referred to as the Hecht equation.

For a MIP particle the situation is different compared to the example described before, as charges are generated over the entire thickness d of the detector. Each moving charge carrier again induces a mirror charge on the detector contact and the total induced contact charge $Q_{ind,MIP}$ by all generated moving charges from the MIP is obtained by integrating equation 6.3 from 0 to d:

$$Q_{ind,MIP} = Q(0)\frac{\mu E \tau}{d}\int_0^d (1 - \exp(-\frac{x}{\mu E \tau}))dx \quad (6.4)$$

The integration leads to the following result:

$$Q_{ind,MIP} = Q(0)\frac{\mu E \tau}{d}\left[1 - \frac{\mu E \tau}{d}(1 - \exp(-\frac{d}{\mu E \tau}))\right] \quad (6.5)$$

The obtained formula is a simplification, as only the drift of one charge type was taken into account. To obtain the generalised equation, a sum over the electron and hole drift has to be performed. The final physical model to describe the reduction of collected charge per MIP $Q_{MIP}(\phi)$ with increased particle fluence ϕ is based on the Hecht equations and the charge carrier lifetime degradation and writes as follows:

$$Q_{MIP}(\phi) = Q_{MIP}(0) \sum_{i=e^-,h} \frac{\mu_i E \tau_i(\phi)}{d}\left[1 - (\frac{\mu_i E \tau_i(\phi)}{d})(1 - e^{-\frac{d}{\mu_i E \tau_i(\phi)}})\right] \quad (6.6)$$

where the sum is performed over the electron and hole properties and $Q_{MIP}(0)$ is the collected charge per MIP without radiation damage. The data from the irradiation were fit with this equation and for the mobilities μ_i the measured values from table 4.1 were used, while $\tau_i(\phi)$ is the charge carrier lifetime at a fluence ϕ and can be further parametrised. The charge carrier life time $\tau_i(\phi)$ is inversely proportional to the number of defects $N(\phi)$: $\frac{1}{\tau_i(\phi)} \propto N(\phi)$. On the other hand the number of defects $N(\phi)$ is proportional to the fluence ϕ: $N(\phi) \propto \phi$. One can therefore write:

$$\frac{1}{\tau_i(\phi)} = k\phi + const \quad (6.7)$$

where k is a constant and $const$ describes the initial conditions at $\phi = 0$ and is therefore equal $\frac{1}{\tau_i(0)}$, while k can be written as the derivative of the inverse lifetime with respect to the fluence:

$$k = \frac{d\frac{1}{\tau_i(\phi)}}{d\phi} \quad (6.8)$$

Finally the charge carrier lifetime can be written as:

$$\tau_i(\phi) = \frac{\tau_i(0)}{1 + \tau_i(0) \cdot \phi \frac{d\frac{1}{\tau_i(\phi)}}{d\phi}} \quad (6.9)$$

Equation 6.6 is further used to fit the degradation curves of the radiation hardness measurements.

The distance that a charge travels during its life time τ before being trapped is the Mean Free Path (MFP, λ). The mean free path can be expressed as a function of the mobility, the electric field and the charge carrier lifetime: $\lambda = \mu E \tau$. For simplification, it is assumed in this expression that the MFP of the electrons λ_{e^-} is equal to the one of the holes λ_h. Due to the linear relation between the MFP and τ, equation 6.9 can be rewritten as:

$$\lambda(\phi) = \frac{\lambda(0)}{1 + \lambda(0) \cdot k' \cdot \phi} \tag{6.10}$$

where k' is the damage constant. This radiation damage parametrisation is used by the RD42 collaboration. The MFP can be larger than the actual detector thickness, in contrary to the Charge Collection Distance (CCD), corresponding to the distance that the charge drifts before being trapped or collected. Hence the maximum of the CCD is the detector thickness d. The relation between CCD, λ and the charge Q_{MIP} is [95]:

$$\frac{CCD}{d} = \frac{Q_{MIP}(\phi)}{Q_{MIP}(0)} = \frac{\lambda}{d} \cdot [1 - \frac{\lambda}{d}(1 - e^{-\frac{d}{\lambda}})] \tag{6.11}$$

6.2 Irradiation facility and beam properties

The irradiation facility is located in the PS East Hall at CERN. It is frequently in use for sample irradiation and detector performance tests [98]. A Beam Position Monitor (BPM) was installed, which measures the beam intensity, the beam profile, the beam position and the beam spread at all time and for each spill as shown for an example cycle in figure 6.2. A local deviation of the beam was immediately visible and alarm messages were sent in case of critical deviations or complete absence of the beam. The fluence and the dose were well controlled parameters during the irradiation [99].

The irradiation conditions were:

- Beam coming from the PS and consisting of protons,
- Particle momentum of 24 GeV/c,
- Beam spread with a FWHM of 1.2 cm at the cryostat and
- Average beam intensity per spill of $1.3 \cdot 10^{11}$ protons/cm^2, corresponding to $1 \cdot 10^{10}$ protons/s on the detectors.

6.3 Measurements for radiation hardness characterisation

In order to characterise radiation hardness, the relevant detector properties (e. g. $Q_{MIP}(\phi)$) have to be measured regularly or continuously. This section discusses the possible methods to measure the property changes of the detector taking into account the conditions of the irradiation in liquid helium.

The challenge of the experiment is to irradiate the detectors, measure their properties and conclude on their radiation hardness, while they are immersed inside liquid helium at all time. In ideal irradiation experiments the procedure is to irradiate the detectors up to a pre-defined total number of particles. Once the wanted fluence is reached, the detectors are taken to the

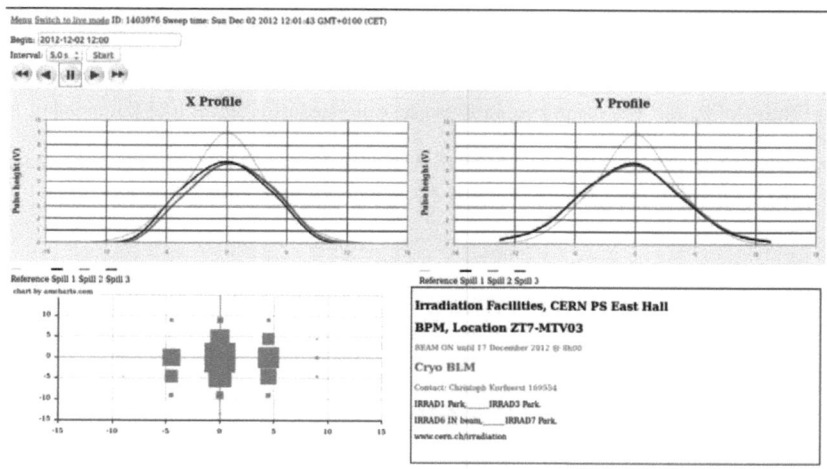

Figure 6.2: Example BPM signals from the December 2012 exposure [100]. The spill intensity in this phase of irradiation was chosen to be lower than the one of a reference spill.

laboratory, where their properties can be precisely measured using a source for example. After these measurements, the irradiated samples can be installed back into the beam for further irradiation and further measurement points. This is not feasible in the case of an irradiation under cryogenic conditions. The aim of the experiment was therefore to monitor the radiation effects during the irradiation itself, without any additional manipulation from outside. The available options were:

1. Single particle detection and measurement of the energy loss distribution of MIPs,
2. TCT (Transient Current Technique see section 4.1) measurements of the current pulse response using a pulsed laser for silicon material,
3. TCT measurements using an alpha source for diamond material and
4. Direct measurement of the current generated by the beam inside the detector samples

These options are elaborated below.

6.3.1 Option 1: Energy loss measurements

From the beam a particle rate of about 400 protons/ns is obtained. The T7 beam is optimised for maximum spill intensity and not for spill intensity variation. Only with careful preparation and tests of the PS operators an intensity decrease of a factor of 7 is possible. At cryogenic temperatures the FWHM of the average pulse response of the tested silicon and diamond detectors to a MIP is below 5 ns, as shown in section 5.4.3. In order to enable such measurements, guaranteeing that each detected event is made out of one particle only, the beam would need to be shifted with respect to the detectors to measure protons coming from the tail of the Gaussian beam. In addition each detector would need to be equipped with an amplifier, placed as close as possible to the detectors, i. e. outside of the cryostat and inside of the radiation area. One downside is the permanent manipulations of the beam, but also the standard CB-50 cables of 50 m length(from the control room to the radiation zone), reducing the quality of the signal, especially with respect to amplitude and rising-falling times of the pulse response. In case of a failure of one amplifier, the contact to the detector would be lost and an exchange inside the zone would need at least one day, due to cooling down period of the activated area after irradiation. This point needs to be further highlighted as in addition the radiation hardness of the amplifier itself is not exactly known.

Previous tests for single particle detection in a cryostat shown in Chapter 5, have shown that these measurements are challenging in two aspects. Not only is the 1.5 m cable between detector inside the cryostat and amplifier outside not ideal, but also the lower part of the Landau distribution overlapped with the observed noise level, hiding the important MPV of the distribution and affecting the determination of the mean value. Under irradiation the charge yield per MIP would become even worse and would make final conclusions very difficult and biased by the measurement method. Therefore this measurement scheme was rejected.

6.3.2 Option 2: Silicon detector TCT using a pulsed laser

The Transient Current Technique (TCT) is a well known method to investigate and assemble the main detector properties, mostly focused on the charge carrier transport phenomena inside the bulk of the detector. For Silicon a 680 nm pico second laser can be used to generate carriers near the surface of the diode. Measuring the TCT signal at different fluencies enables to monitor the changes in the detector bulk induced by irradiation. In order to bring the laser light from the control room to the detectors, a special setup with radiation hard single mode optical fibres was designed with the help of Elisa Guillermain.

6.3.3 Option 3: Diamond detector TCT using an alpha-particle source

Due to the large electron hole pair creation energy of 13 eV for diamond material, 680 nm laser light can not generate carriers. For diamond detectors an alpha source is needed. Unfortunately the TCT measurements with alpha particles on diamond during cryogenic irradiation are difficult because:

- Even non-irradiated sCVD show a factor of 2.8 smaller signal at liquid helium temperatures compared to room temperature (see figure 4.18),

- Radiation induced defects will further reduce the signal,

- Significant background is expected from activated material (betas, gammas), strongly reducing the quality of the measurements and the reliability of triggering on events from alphas only,

- Effect of increased polarisation under irradiation was difficult to estimate and could lead to no alpha-particle signal at all,

- No implementation of an external trigger is possible and

- No alpha source can be used in cryogenic conditions and under irradiation due to radiation protection issues. The option of gluing a material on the detector, which would under irradiation become an alpha source was investigated, unfortunately without success, as all materials would produce further unwanted radiation due to their activation.

It was therefore decided that such a measurement is not feasible for the diamond detectors. This could be solved in future measurements through the use of a short-wavelength laser able to generate charges in the diamond material, instead of the alpha source.

6.3.4 Option 4: DC measurements

DC measurements have the advantages of being highly fail-safe, which is a major criteria with respect to the complexity of the task and the non-accessibility of the detectors. In addition the measurements allow a continuous monitoring of the leakage current. Also polarisation effects can be directly observed and in the best possible way counteracted through programmed voltage modulation. The further important advantage is the evaluation of the detectors radiation hardness with respect to DC measurements, which is the preferred readout for the final BLM application.

The RD42 collaboration used DC read-outs for detector's radiation hardness measurements with success in 1999 already [97].

The drawbacks of the DC measurement method are:

- Missing of the exact initial values measured with beam (before irradiation), because especially at the beginning the reduction in detector's signal is most significant (well visible in linear plots of the signal degradation compared to double logarithmic plots, as shown in section 6.5.2) and first spills were needed to align the beam and check the correct functioning of all detectors and instruments,

- High charge carrier density within the detector bulk during the spill possibly leading to regions with a disturbed electric field. This is expected to be less noticeable for diamond due to the larger band gap (smaller recombination probability) and at cryogenic temperatures, because of higher charge carrier drift mobility,

- Strong dependence of the quality of the measurements on beam stability - in principle it is possible to correct for each change, like beam spread variations and temporary slight misalignments. The argument is that with the large number of points, the beam variations cancel out, hoping for small systematic effects/errors in beam properties. Anyhow the direct comparison of the detectors is perfectly valid, as they get the same beam at all time.

6.3.5 Conclusion on measurement options for radiation hardness characterisation

After careful comparison of all advantages and disadvantages together with the risk assessment of operation in a non-accessible liquid helium and irradiation environment, the two selected measurement options were the laser TCT on silicon detectors and the continuous DC measurements on all detectors. A first testing of the DC measurements at room temperature with a silicon detector and a diamond detector was possible before the more challenging tests in liquid helium. These irradiation tests will be presented.

6.4 Radiation hardness at room temperature

The room temperature measurements were performed during August, September and October 2012 at CERN in the T7 beam line of the East Area. The goal was to test if the characterisation of the detector's radiation hardness is feasible with the proposed method. In addition it could be used to check if the designed readout chain works correctly and to measure its noise and frequency characteristics.

6.4.1 Experimental setup

Detector modules

Two CIVIDEC holders for DC measurements (example picture shown in figure 6.3) were installed in the beam.

Figure 6.3: CIVIDEC detector holders for irradiation and DC measurements.

The DC setup had two cables, one was for signal readout and the other for voltage application.

Due to the proximity of high current and high field magnets, no magnetic material was used for the modules and the supports.

Signal readout and data acquisition

Direct Current (DC) measurements were done using a Keithley 6517. The instrument allowed to apply the voltage and measure the detector current. For the data acquisition a LabVieW program was written. The NPLC (number of power line cycles) of the current measurements was set to one, corresponding to 16.67 ms, during which the recorded current value I was averaged. The current measurements were not synchronised with the spill. Therefore the measurement points were random with respect to possible coherent variations within a spill, so that through the averaging over several spills a correct spill reconstruction was done.

In the offline analysis the current I was integrated over the spill duration and the obtained total charge was normalised by the number of particles. The obtained normalised charge for the spills of one series of measurements were then averaged. The analysis procedure is summarised in equation 6.12. The spill intensity Np was known from a Secondary Emission Chamber (SEC) [99]. In addition to the SEC value, Figure 6.4 shows the current measurement of a detector in the beam, together with the measurement value from the SEC.

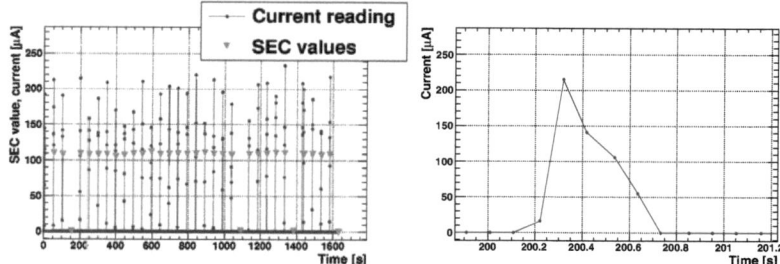

Figure 6.4: Example measurement series with a diamond detector. Each peak in the left plot corresponds to a beam induced detector current. The green triangles are the values measured with the SEC. They are proportional to the number of protons in one spill. The right plot is a zoom into the current measurement during one spill.

$$\overline{Q}_{MIP} = \frac{1}{n}\sum_{i=1}^{n} \frac{\sum_j (I_{ij} - Ioff_i)(t_{ij} - t_{ij-1})}{Np_i} \qquad (6.12)$$

where:

- $\overline{Q}_{MIP}$ is the mean collected charge per MIP,
- n is the number of measured spills of one series of measurements,
- I_{ij} is the j-th over 16.67 ms averaged current measurement of the i-th spill,
- $Ioff_i$ is the mean measured offset current before the arrival of the i-th spill, corresponding to e. g. leakage or forward currents,
- t_{ij} is the time of the measurement I_{ij},
- Np_i is the total number of protons of the i-th spill measured by the SEC.

The measurement range was adapted according to the signal amplitude, which changed with fluence, beam intensity, detector material, applied voltage and temperature. For the measurement of the current-voltage characteristics without beam, the range was set to as low as 20 pA at 1.9 K and up to 2 mA in case of irradiated silicon at room temperature.

Detector samples

For the RT irradiation tests, a p^+-n-n^+ silicon detector of 10 kΩcm resistivity with aluminum metallisation was used. The thickness of the sample was of 300 μm.

In addition one single crystal chemical vapour deposition diamond detector (sCVD) with a double layer metallisation of titanium and gold and a thickness of 500 μm was irradiated.

The detector characteristics are summarised in table 6.1.

Detector	Thickness [μm]	Active area [mm^2]	Metallisation
Si-10kOhm-T9	300	7x7	Al
sCVD-T9-1	500	4.7x4.7	Ti + Au

Table 6.1: Detector samples for the measurements at room temperature in the high intensity beam. The detectors were used in the low intensity beam test already (see Chapter 5), where they measured an integrated number of 9 GeV/c protons below $3 \cdot 10^9$ cm^{-2}. This number is negligible with respect to the number of protons ($1.3 \cdot 10^{11}$ cm^{-2}) in one single spill of the high intensity beam.

6.4.2 Experimental results

A total integrated fluence of $6 \cdot 10^{15}$ protons/cm^2 for the diamond detector was reached at the end of the room temperature irradiation, corresponding to an integrated dose of 1.7 MGy for the diamond sensor. The mean rate of energy loss $-\langle dE/dx \rangle$ calculated with the Bethe-Bloch equation in section 5.1 was used to estimate the dose D in the tested particle detectors during the irradiation:

$$D = n \frac{-\langle dE/dx \rangle}{\rho} \tag{6.13}$$

where n is the number of particles. The dose deposited by a MIP in silicon material hence corresponds to 2.67 nGy/cm^2, while for diamond material it is of 2.82 nGy/cm^2.

As with increased irradiation the concentration of radiation induced traps increases, the internal electric fields inside the detector bulk might be non-uniform. It is therefore preferred to specify the external applied voltage during the measurements instead of the electric field. The mathematical equality of the two notations is given, as the integral of the electric field over the detector thickness is equal to the applied voltage, regardless of the internal inhomogeneities.

For silicon not more than 400 V reverse bias was applied as the leakage current increased under irradiation at 303 K.

Leakage current

Before irradiation the leakage current of the silicon detector was of 45 nA at 100 V, while for sCVD the leakage was below 50 pA at 500 V.

During the irradiation the leakage current of silicon increased as shown in figure 6.5. The initial increase was linear, which is shown as the blue line in the plot. After a certain fluence, a saturation of the leakage current was observed, which was fitted with an exponential function shown in green. The saturation effect was due to the fact that the silicon detector was not fully depleted under the application of 100 V. Not all regions of the detector saw an electric field and therefore not the entire detector bulk contributed to the leakage current after a certain fluence. Under the application of higher voltages the linearity is expected to be observed until higher fluences. This linearity of the leakage current with fluence has been described in detail in [101].

After the irradiation the measured IV curves can be seen in figures 6.6.

The diamond detector leakage currents shown in figure 6.6 stayed low for all voltages, even after high irradiation. This is one of the main advantages of diamond sensors compared to silicon detectors at ambient temperatures. After irradiation the silicon material leakage current was of 48 µA at 100 V. As already mentioned in earlier sections, this major disadvantage disappears at cryogenic temperatures, even under high irradiation.

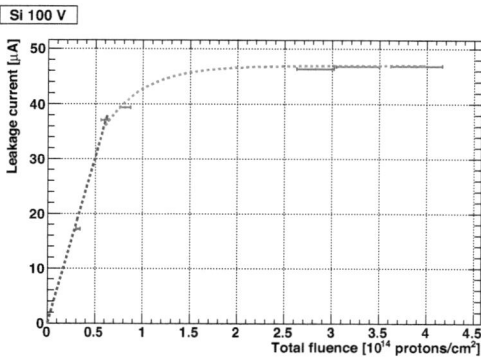

Figure 6.5: Silicon detector leakage current during irradiation.

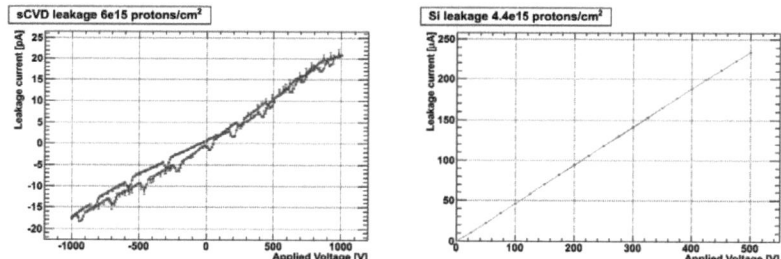

Figure 6.6: Diamond and silicon detector leakage curves after irradiation. The leakage current of silicon is about 7 orders of magnitude larger compared to the diamond detector. Before irradiation the reverse current of the silicon sensor was of 45 nA at 100 V, while for the diamond detector the measured dark current was below 50 pA at 500 V.

Degradation curves

In figure 6.7, the signal decrease with increased proton fluence is shown for the diamond and the silicon detectors. The fit yields an initial collected charge between 2 and 3 fC per MIP for the diamond sample. The colour code shows how with higher voltages more charges can be collected.

The use of silicon detectors at 303 K, under irradiation and in DC mode was unfavourable in comparison to the diamond detectors. The leakage current increase was immediate, while for the diamond material the leakage current stayed under 100 pA even under high irradiation. A radiation hardness comparison of the two detectors using DC measurements is therefore not reliable and no final conclusions should be drawn from these measurements concerning the radiation hardness of the silicon detector. In addition the high charge carrier density effect (plasma effect) was noticeable in case of the silicon sensor, because the initial collected charge was below the theoretical value at 100 V, although the detector should be fully depleted. With higher applied voltages the theoretical value was reached. The generation of charges from the beam spill lead to a charge density of $\approx 5 \cdot 10^{13}$ charges/cm^3, which lead to perturbations of

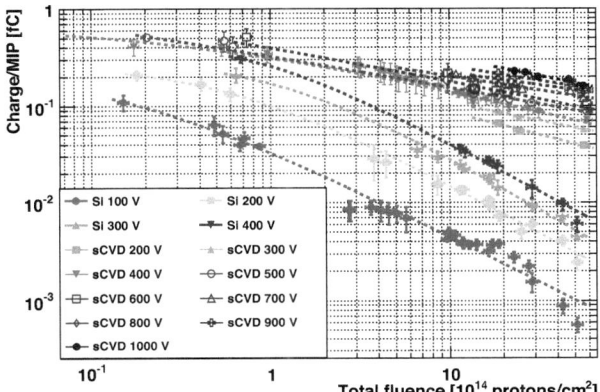

Figure 6.7: Degradation curves of the single crystal CVD diamond detector and the silicon detector at 303 K. The model based on charge carrier lifetime degradation and Hecht equations given in equation 6.6 is used to fit the data and is shown in the graph as the dotted lines.

the electric field distribution within the detector bulk. In silicon material at room temperature this is pronounced by the lower band gap and the slower charge mobility compared to diamond material and compared to silicon detectors at cryogenic temperatures.

The diamond detector data from the T7 beam test could not be adequately fit with the equations 6.6 and 6.10. The reason was that the measurements showed a fluence dependence of the damage parameter $k \propto \phi^{s-1}$, instead of a constant k. The parameter s was introduced to mathematically take the observed deviation from the expected fits into account. It was further chosen to keep k a constant and instead introduce the parameter s as an exponent to the fluence ϕ. The complete fitting equation is hence:

$$Q_{MIP}(\phi) = Q_{MIP}(0) \sum_{i=e^-,h} \frac{\mu_i E \tau_i(\phi)}{d} \left[1 - (\frac{\mu_i E \tau_i(\phi)}{d})(1 - e^{-\frac{d}{\mu_i E \tau_i(\phi)}}) \right] \quad (6.14)$$

with the charge carrier life time τ_i, the damage constant k and the exponent s:

$$\tau_i(\phi) = \frac{\tau_i(0)}{1 + \tau_i(0) \cdot k \phi^s} \quad (6.15)$$

The introduction of the parameter s has shown to largely improve the quality of the DC-measurements fit for all tested detectors, all applied voltages and the experiment temperatures. An example of this is shown in figure 6.8. Its physical origin and explanation is not yet clear. The consequence of $s \neq 1$ was that there was an additional unknown effect proportional to ϕ^{s-1} leading to a charge lifetime degradation not proportional to the fluence. This was also observed during the cryogenic irradiation and further discussions will follow in section 6.5.2.

Voltage Scan

The voltage scans for the silicon and diamond detectors at different fluences are depicted in the figures 6.9. In non-irradiated single crystal CVD diamond detectors a full charge collection is

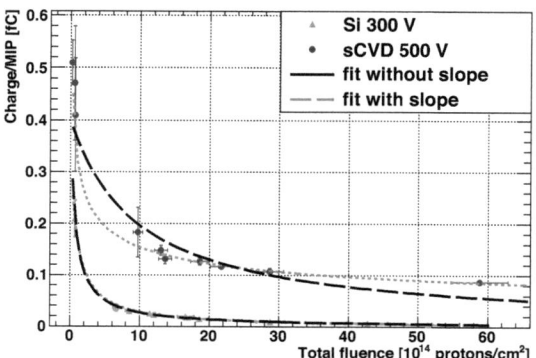

Figure 6.8: Visualisation of the effect that the introduction of the parameter s has on the fit of the data. In case of silicon material at room temperature the two fits are very similar, as the best fit yields a slope s equal to 1. The normalised χ^2 of the fit for the diamond detector with slope s is of 0.6, while in the case without slope it is of 6.3.

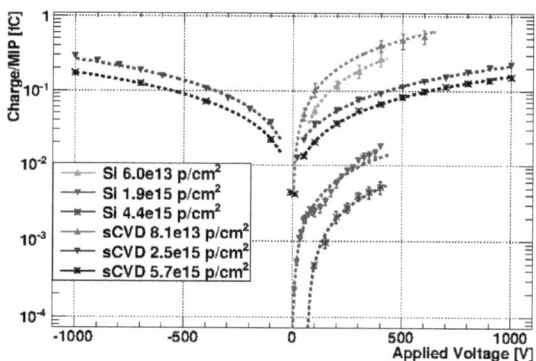

Figure 6.9: Voltage scan for sCVD and for 10 kΩcm silicon. The y-axis is logarithmic.

reached at a voltage of 100 V. This was measured in the low intensity beam test and was shown with alpha particles in [70]. For the irradiated sample, there was no full charge collection at 100 V. This is similar for the silicon detector, where an initial full depletion voltage did not lead to a full charge collection due to the radiation defects. For sCVD a consistent asymmetry for positive and negative voltage of about 10 % was observed. This was measured by other groups in the RD42 collaboration as well (ETH Zürich and M. Guthoff). One probable explanation for this observation is that the asymmetry is due to differences between the two contacts of the diamond detector. The shown fitting is linear and describes the data well, as no signal saturation was reached within the range of the applied voltages.

The slope from the linear fit of the voltage scan is plotted in figure 6.10 versus the particle

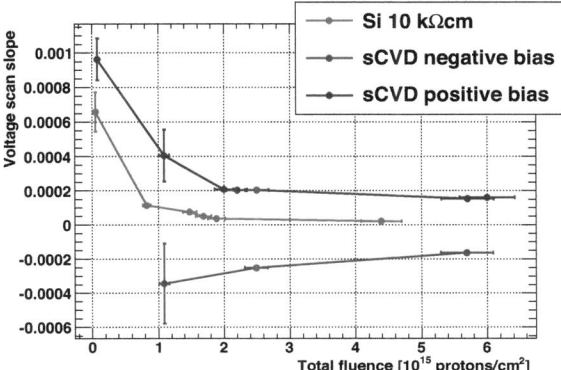

Figure 6.10: Slopes from voltage scan fits plotted versus the fluence. The lines between the data points are for visualisation only.

fluence. It is a measure for the potential of the detector to increase the charge collection efficiency through the application of higher voltages.

The dependence of the fitting parameters on the applied voltage are shown in figure 6.11. For the silicon and diamond detectors, the damage constant k could be decreased through the application of higher voltages. For the parameter s, the changes with voltages were not significant for both detectors, making it a detector constant. For the silicon detector, the parameter s was equal to one for all measured voltages at room temperature, which is in agreement with the expectations.

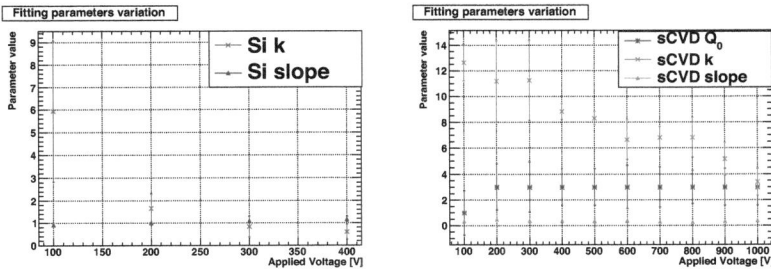

Figure 6.11: Fitting parameter variations with applied voltage for the silicon and diamond detectors. Q_0 is in [fC].

6.5 Radiation hardness at liquid helium temperatures

The detector performance at liquid helium temperatures under irradiation was unknown. One expectation for silicon sensors was that the leakage current decreases at 1.9 K even under high irradiation, hence ameliorating the detector performance. Another expectation was that no de-trapping of charges, even from shallow defects, occurs at liquid helium temperatures. This could have a positive effect on the detector signal, as the occupied radiation defects are neutralised and can not trap further charges, but it can also have a negative effect, because stably trapped charges lead to possible detector bulk polarisation and inhomogeneous electric fields. In addition a possible accumulation of charges between the detector material and the metallic contacts could highly deteriorate the detector performance. Therefore even a complete disappearance of the detector signal was conceivable. Only irradiation tests could bring the final answer about the actual detector radiation hardness in the cold.

During November and December 2012 the irradiation in liquid helium in the T7 beam line at CERN could be performed and is summarised below.

6.5.1 Experimental setup

Figure 6.12 shows the inside of the irradiation zone with installed equipment. The cryogenic components were especially designed and manufactured for this experiment.

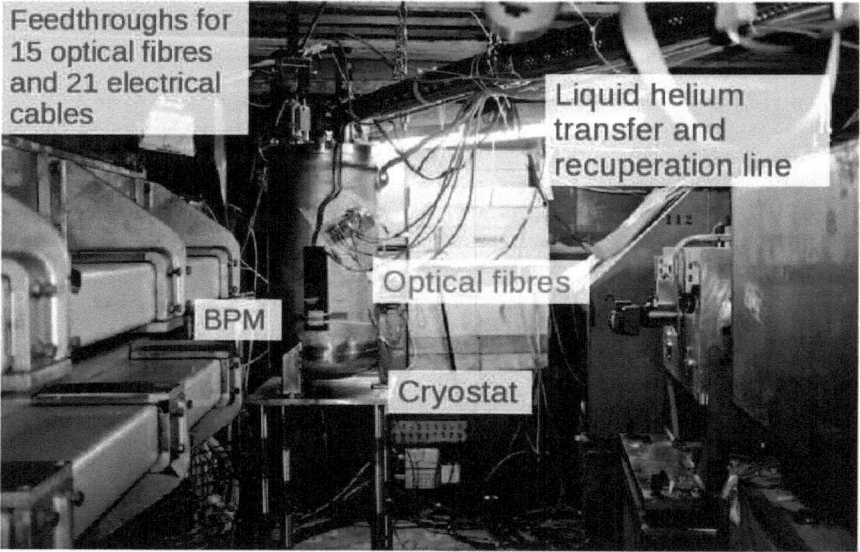

Figure 6.12: Cryostat at its final position in the irradiation area. Each element visible in the picture was especially constructed for this irradiation test.

Cryogenic system

The cryogenic system was adapted to match the challenging requirements of the radiation test facility. The main elements of the cryogenic system were the cryostat, the helium storage Dewar, the transfer line that connected the two of them and the vacuum pump as seen in figure 6.13. The helium dewar is a storage tank capable of holding up to 500 litres of liquid helium at 4.2 K. The cryostat and the transfer line are also superinsulated, so that the helium could stay at a liquid state during transfer and the boil-off rate of the helium within the cryostat is as low as possible. The in-house produced transfer line supplies liquid helium to the cryostat and transports the cold gas out of the radiated area. It consists of two concentric flexible conducts which are vacuum-jacketed and lined with multi-layer insulation (MLI). This type of insulation was often used in cryogenics as it reduces dramatically the heat load from the environment. The length of the transfer line is 12 m, and it spans the distance between the cryostat, located in the radiation area, and the dewar, which remained outside the beam zone with the rest of the equipment, such as the primary pump and the helium gas bottle.

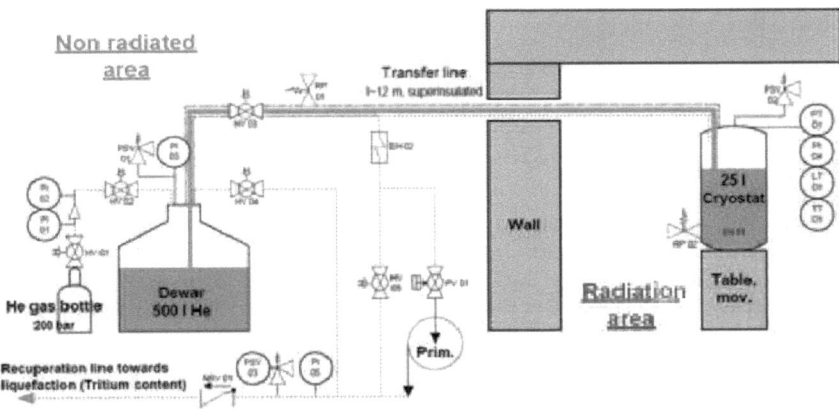

Figure 6.13: Cryogenic process and instrumentation diagram. Courtesy of Thomas Eisel.

The transfer of liquid helium from the storage dewar to the cryostat was performed by creating a slight overpressure in the dewar, i.e. pressurising the dewar to 200 mbar with helium gas, supplied by a standard pressurised gas bottle.

After an approximate volume of 30 litres of liquid helium had been reached within the cryostat the transfer was stopped by equalising the pressures in the dewar and the cryostat. This was achieved by opening the dewar back to the venting line that connects both the dewar and the cryostat to the helium recuperation system, which had a pressure slightly higher than the atmospheric pressure to avoid air leaking into the cryogenic system. The temperature of the helium bath was then lowered from 4.2 K to 1.9 K by pumping away the vapour space above the liquid line as described in section 4.2. During this process, the helium, initially a saturated liquid at a pressure of 1 bar and 4.2 K, followed the saturation line. The pumping of the bath was carried out by a primary pump. The pump outlet was connected to the recuperation system. The saturation pressure in the helium bath during the pumping was controlled by a metering pump that automatically regulates the flow through the pump. Further information on the details of the cryostat design, the transfer line design, the safety regulations concerning the cryogenic equipment and the functionalities of all implemented cryogenic instrumentation can be found in [102].

The temperature of the samples during irradiation was between 1.9 K and 4.2 K. With the used measurement method, no significant difference of the detector properties with respect to timing and charge collection efficiency within this temperature range was observed. The cryostat was refilled two times per day, to guarantee a stable temperature over the entire duration of the irradiation. The average duration of a refill cycle took about 80 minutes.

Towards the end of the irradiation, one warm up cycle up to 80 K was performed, to simulate the LHC shut-down period over Christmas, during which the cold mass of the magnets is warmed up to this temperature. The Goal was to see what effect this has on the detector performance and if annealing can be observed. Finally the last two days of irradiation were used to warm the cryostat up to room temperature and measure the temperature dependence of the detector properties.

Detector holder

Three different detector holder types were used in the experiment:

- 3 holders for DC measurements (example picture shown in figure 6.14)
- 5 holders for TCT measurements (example picture shown right of figure 6.15)
- 2 holders with 4 silicon detectors each, as telescope (example picture shown left of figure 6.15)

The DC holder had two cryogenic coaxial UT 85 cables, for low heat introduction. One was for signal readout and the other for voltage application.

The TCT setup enabled the installation of two single mode optical fibres, one for each detector side. The laser light was then guided with aluminum mirrors on the two detector sides. Electrically the n^+ side of the silicon detector was connected to ground, while the p^+ detector side was used to apply voltage and read signal at the same time.

Figure 6.16 shows all modules arranged on the ground plate and ready to be placed inside the cryostat. Aluminum foils were attached as dosimeters at the outer extremities of the detector holders, to confirm the total dose at the end of the irradiation.

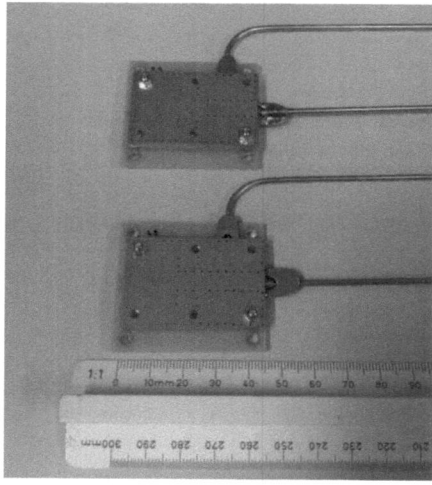

Figure 6.14: Detector holders from CIVIDEC for DC readout.

Due to the proximity of high field magnets, no magnetic material was used for the detector holders and the supports.

Figure 6.15: In the left picture is shown the silicon detector telescope from the Ioffe institute in Russia used for the verification of the correct alignment with the beam. In the right picture are shown detector holders from the Ioffe institute in Russia used for the laser TCT measurements.

Signal readout and data acquisition

During the irradiation three different readouts were used.

DC readout

The DC measurements of the current generated by the beam in the detector was measured in the same way as at room temperature, see section 6.4.1. DC measurements were also performed for the detectors in TCT holders. This was achieved through the use of a Keithley 2410 electrometer to read the currents produced in the silicon detectors by the beam particles.

TCT readout

The TCT measurement readout was done using a 20 dB amplifier from CIVIDEC to amplify the signal coming from the silicon samples. The acquisition of the pulse response was done with a 3 GHz LeCroy oscilloscope. The picosecond laser light was generated by a PiLas Digital Control Unit (EIG1000D) and an optical head for 680 nm.

Telescope readout

All 8 silicon detectors from the two telescopes were connected in parallel to the same voltage on the n^+ side. The p^+ side was directly connected through standard CB-50 cables to a 1 GHz LeCroy oscilloscope wih 50 Ω DC input impedance, to visualise the DC voltage variations due to the beam particles.

Thus the spill shape was measured by each quadrant of the telescope detectors. The criterion for correct alignment of the detectors relative to the beam is a symmetric distribution of the current between the quadrants in both the input and output telescope detectors.

Figure 6.16: Detector modules mounted on the support plate and ready for cooling down and irradiating.

Detector samples

For the irradiation tests p^+-n-n^+ silicon detectors of different resistivity have been used. The investigated samples were devices of 4.5 Ωcm, 200 Ωcm, 500 Ωcm and 10 kΩcm with Al metallisation. The thickness of the samples was of 300 µm.

In addition 2 diamond detectors have been used, both of them being single crystal chemical vapour deposition (sCVD) diamond detectors with a double layer metallisation of gold and titanium and a thickness of 500 µm.

The detector samples under investigation are summarised in table 6.2.

Detector	Resistivity [Ωcm]	Thickness [µm]	Active area [mm^2]	Metallisation
Si-10kOhm-Telescope × 8	10k	300	10×10	Al
Si-10kOhm-T7-1	10k	300	5×5	Al
Si-10kOhm-T7-2	10k	300	5×5	Al
Si-10kOhm-T7-small	10k	300	3×3	Al
Si-500Ohm-T7	500	300	5×5	Al
Si-4.5Ohm-T7	4.5	300	5×5	Al
sCVD-T7-1		500	4.7×4.7	Ti + Au
sCVD-T7-2		500	4.7×4.7	Ti + Au

Table 6.2: Detector samples for the cryogenic measurements in the high intensity beam.

Beam alignment system and procedure

High importance was given to the mechanical securing of a best possible alignment of the detectors inside the cryostat with respect to reference points on the outside of the cryostat. The length contractions of the used materials caused by the temperature difference when cooling with liquid helium were taken into account.

A Beam Position Monitor (BPM) was installed outside the cryostat at the placement of the detectors. A first alignment after installation of the cryostat in the irradiation zone was done visually with a laser, aiming at the centre of the BPM.

In addition a silicon telescope (shown in figure 6.15) at the outer positions of the detectors inside the cryostat allowed verifying the alignment with respect to the BPM on the outside. The telescope modules contained 4 silicon detectors each.

6.5.2 Experimental results

At the end of the irradiation a total integrated fluence of $1.22 \cdot 10^{16}$ protons/cm^2 was reached, corresponding to an integrated dose of 3.26 MGy for silicon material and 3.42 MGy for the diamond detectors. For the silicon detectors at liquid helium temperatures, measurements could be performed under forward bias application, which is known as Current Injected Detector (CID) [39] and was described in section 4.3.3.

The indication of the external applied voltage is further preferred, compared to the possibly miss-leading electric field. This is because with increased irradiation the number of charge traps increases, which may lead to regions of inhomogeneous electric fields. This is pronounced by the fact that at liquid helium temperatures no charge de-trapping is expected. It should be mentioned that the mathematical equality of the two notations is given, as the integral of the electric field over the detector thickness is equal to the applied voltage, regardless of the internal inhomogeneities.

Leakage current

Figures 6.17 show that not only the reverse silicon leakage current goes down to 50 pA at 100 V, but also the forward current is only of 60 pA at -400 V for an irradiated silicon detector in cold.

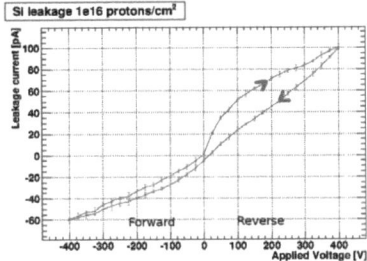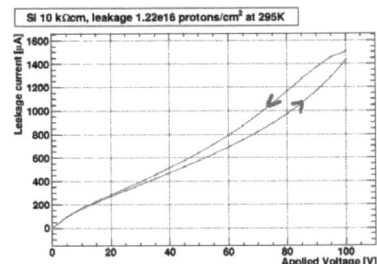

Figure 6.17: Silicon leakage curves. On the left side is the leakage current at liquid helium temperatures, while on the right side is the leakage current at room temperature after the irradiation. Positive voltage denotes reverse bias and negative voltage corresponds to forward bias.

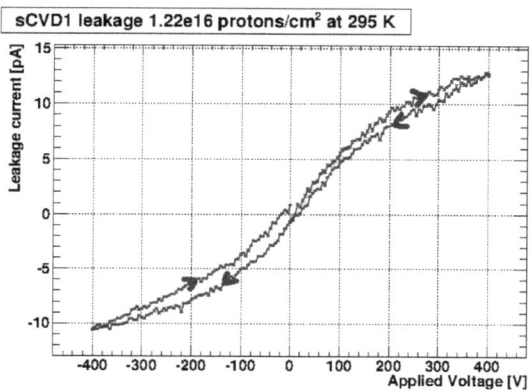

Figure 6.18: Diamond leakage curve after irradiation at 295 K.

The diamond leakage current seen in figure 6.18 stayed low for all voltages and also for temperatures up to room temperature, even after high irradiation, which is one of the main advantages of diamond compared to silicon at ambient temperatures.

The beam had to be stopped for the leakage current measurements. Therefore only a limited amount of time was dedicated to it. The observed hysteresis phenomena was an effect partly due to this time restriction. The leakage current relaxation was related to the time constant $\tau = RC$, where R is the resistance and C the capacitance of the detector. After the voltage application, the leakage current tended towards its stable value with an exponential time constant of 320 s at 1.9 K for the diamond sample. At room temperature the stabilisation time is shorter.

Degradation curves

Figure 6.19 shows the signal decrease with increased proton fluence for the single crystal CVD diamond detector. The curve for the 500 Ωcm silicon sensor with 100 V reverse bias has been plotted in all graphics as reference curve.

The figures 6.20, 6.21, 6.22 and 6.23 show the reduction in charge collection for the silicon devices. The fits were performed with the same formula as at room temperature and again the introduction of the parameter s lead to a major improvement of the quality of the fits (χ^2).

For non-irradiated silicon detectors and at a low radiation dose, the forward current is of about 100 μA at 100 V in liquid helium. With increased radiation damage, the forward current decreased and becomes as low as shown in section 6.5.2. This was one reason why early points were missing for forward bias silicon. Further observations were that the forward bias for low resistivity silicon sensors is less stable than for high resistivity silicon detectors. This was especially visible for the 4.5 Ωcm sample, which underwent a jump in its charge collection under forward bias operation, as shown in figure 6.23. The charge per MIP of the 4.5 Ωcm resistivity silicon sample was initially stable and low and reached the levels of high resistivity detectors after the jump. This will later also be visible during the performed voltage scans in section 6.5.2. Further to mention with respect to its instability, is the observation that its leakage current increased after the warm-up cycle to 80 K and stayed unstable with

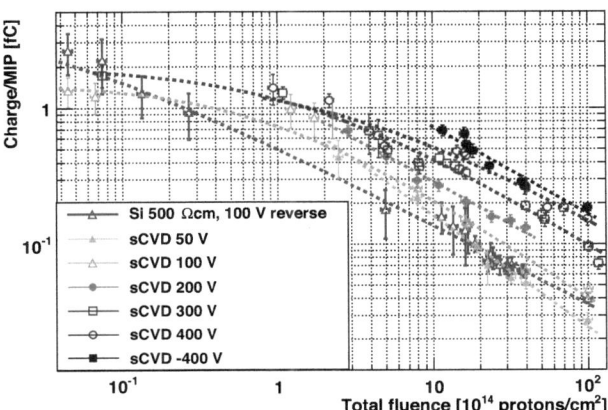

Figure 6.19: Degradation curves of the single crystal diamond detector at different voltages with the silicon 500 Ωcm 100 V reverse as reference curve.

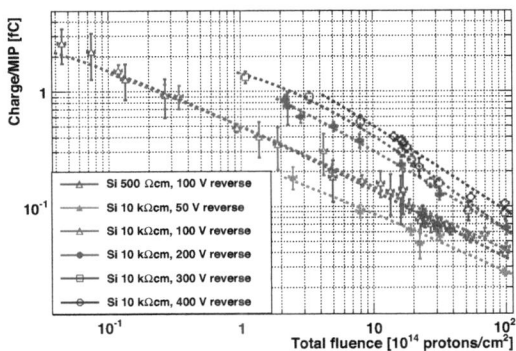

Figure 6.20: Silicon 10 kΩcm degradation curves under reverse bias with the "silicon 500 Ωcm 100 V reverse" as reference curve. The 100 V reverse curves of the two detectors with different resistivity match each other.

non-systematic variations until the end of the irradiation.

To determine the quality of the fit, a distribution of the normalised χ^2 has been produced and is shown in figure 6.24. The mean value of the distribution is of 1.2. The normalised χ^2 smaller than one is a sign for an overestimation of the errors. Normalised χ^2 values larger than one indicate that the data is not perfectly described by the fit. A statistical difference between the experiment and the fit is present. This might be due to instabilities of the beam during the measurement or due to effects within a detector that are not yet fully understood. For example at small applied voltages it is rather improbable that the internal field in the

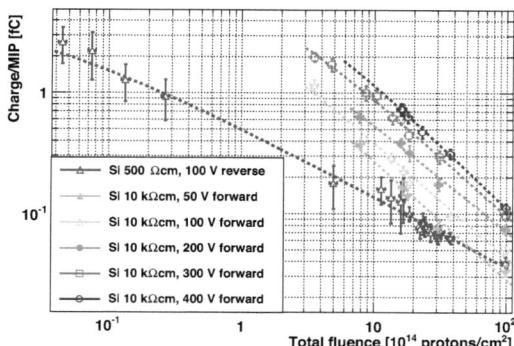

Figure 6.21: Silicon 10 kΩcm degradation curves under forward bias with the "silicon 500 Ωcm 100 V reverse" as reference curve.

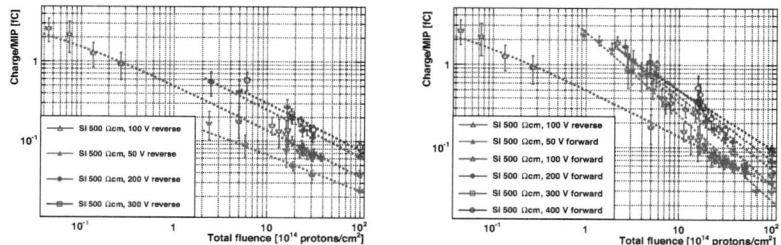

Figure 6.22: Silicon 500 Ωcm degradation curves. The degradation curves under reverse bias are shown in the left plot, while the results under forward bias are depicted in the right plot, together with the "silicon 500 Ωcm 100 V reverse" as reference curve.

detector bulk is the same over the entire irradiation. Overall the proposed fitting formula allows a description of the degradation curves with a normalised $\chi^2 \approx 1$ for all detectors, all measured temperatures and all voltages. Without the introduction of the parameter s in the fit, the mean normalised χ^2 is around 6.9, which is significantly higher.

Parameter s discussion

The introduction of the parameter s lead to a major improvement of the fitting quality as earlier shown for an example degradation curve at room temperature in figure 6.8. Together with the degradation curves at liquid helium, two groups of curves with respect to s can be distinguished:

- those with $s = 1$ (as expected) for the silicon detectors at room temperature and silicon in liquid helium under forward bias

- those with $s < 1$ for the diamond detectors at all temperatures and the silicon detectors in liquid helium under reverse bias.

The group with $s = 1$ had in common that the detectors had a leakage (respectively

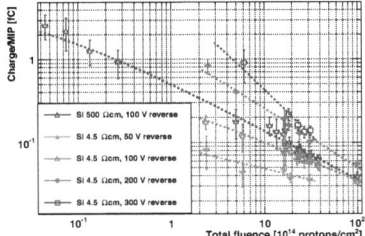 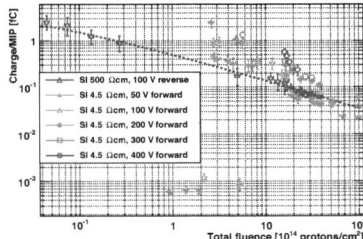

Figure 6.23: Silicon 4.5 Ωcm degradation curves. The degradation curves under reverse bias are shown in the left plot, while the results under forward bias are depicted in the right plot, together with the "silicon 500 Ωcm 100 V reverse" as reference curve. Under forward bias voltage the data points of the 4.5 Ωcm silicon detector sample can not be fit, as the sensor was subject to a sudden increase during which the initial low and stable charge per MIP arrived at a level comparable with the higher resistivity samples.

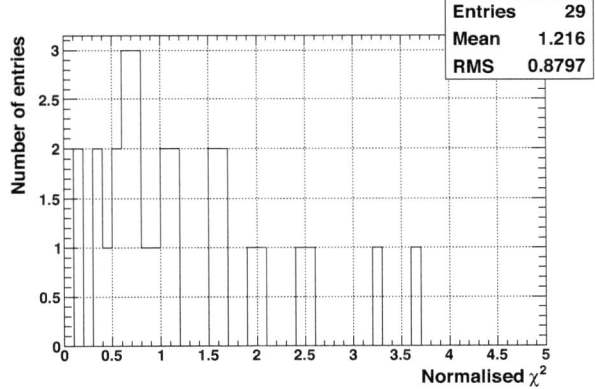

Figure 6.24: The normalised χ^2 distribution is shown in this figure to check the quality of the fits.

forward) current, while for the group with $s < 1$, no leakage currents above 100 pA were measured. The additional effect (proportional to ϕ^{s-1}) could therefore be related to the absence of a leakage current. It should be clearly mentioned that the observation from the measurements was that $s < 1$ when no leakage current is measured, which is not equivalent to a universal causality suggesting that no leakage current leads to $s < 1$. These two statements need to be distinguished.

Silicon and diamond detector comparison

In order to compare silicon and diamond detectors, degradation curves with similar electric fields have been chosen, which means for the silicon sensors an applied voltage of 300 V in reverse and forward modes and for the diamond detector 400 V. Figure 6.25 depicts the degradation curves of the 10 kΩcm silicon detector and the ones of the single crystal diamond detector. The silicon diode had a larger signal than diamond detectors at the beginning of the irradiation, but the situation changed with increased fluence. The forward bias modus for silicon lead to high signals at the beginning of the irradiation, meaning a large $Q(0)$. Unfortunately the signal decrease at higher fluence was faster compared to the reverse bias operation and the diamond detectors.

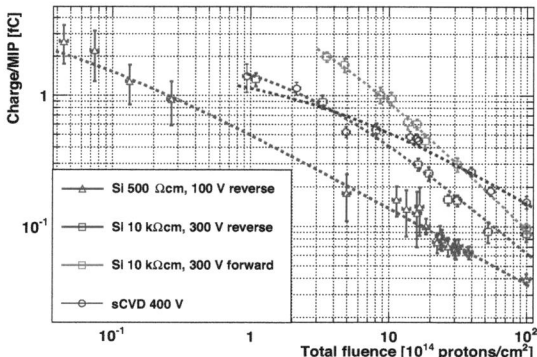

Figure 6.25: Degradation curves of single crystal diamond detector at 400 V compared with 10 kΩcm silicon detector at 300 V and the "silicon 500 Ωcm 100 V reverse" as reference curve. Some measurement points at low dose are missing, due to three main reasons: (1) the alignment procedure at the beginning of the irradiation, (2) the initial measurements of the detectors with an applied voltage of 100 V and (3) the large forward current at low radiation damage for the detectors in CID operation.

At low irradiation dose, silicon detectors operated at 300 V reverse bias had a larger signal than the diamond detector with 400 V bias. The crossing point was at a fluence of $3.8 \cdot 10^{14}$ protons/cm^2 (0.1 MGy), from where on sCVD started to have higher signal.

The crossing point between sCVD with 400 V and the silicon detector with 300 V forward bias was at $3.35 \cdot 10^{15}$ protons/cm^2 (0.9 MGy), showing that for very high radiations diamond sensors should be the material of choice.

A further improvement of the radiation hardness can be obtained through the use of thinner detectors, as charges drifting through the detector are collected earlier and their trapping probability is hence reduced. For thin detectors the mean free path of the charges is larger than the detector thickness until a higher fluence and hence the charge collection distance stays constant until then. This can be understood in detail from equation 6.11. The downside of a thin detector is the measurement of a lower total charge compared to a thick detector. This can be compensated through the use of a stack of detectors with a total thickness equal to a thick detector.

Voltage Scan

The voltage scans for the different detectors at different fluencies are shown in the figures 6.26, 6.27 and 6.28. In the voltage scans for silicon detectors, positive voltage denotes forward bias application.

No full charge collection was given for the single crystal diamond detectors above a voltage of 100 V, as it would be expected for non-irradiated single crystal diamond detectors at room temperature. The situation for silicon detectors was similar, where initial full depletion voltage did not lead to full charge collection any more.

Applying higher voltages, a saturation of the charge collection was expected, but for the considered voltage range up to 400 V in cryogenic conditions, a linear fit is adequate and has been used for the plots and the further analysis.

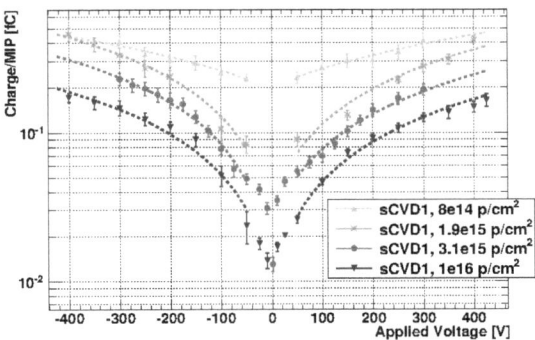

Figure 6.26: Voltage scan for sCVD at different fluences.

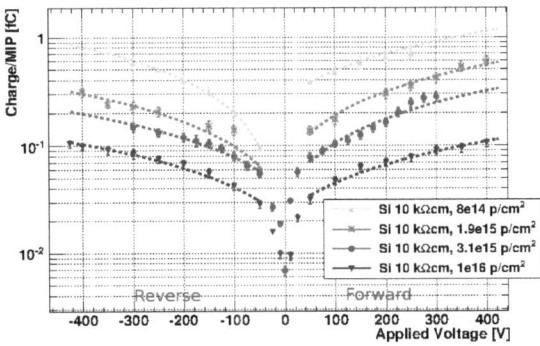

Figure 6.27: Voltage scan for the 10 kΩcm silicon detector at different fluences.

The slope shown in figure 6.29 is a measure for the potential of the detector to increase the charge collection through the application of higher voltages. The peak observed for the single crystal CVD diamond detector suggests the existence of three stages:

1. Initially very small slope for non-irradiated detectors, because a full charge collection was reached for low voltages already,

2. Increase of the slope, because higher voltages were needed for a full charge collection efficiency (visible in the comparison of damage constant k between voltages, shown in figure 6.30)

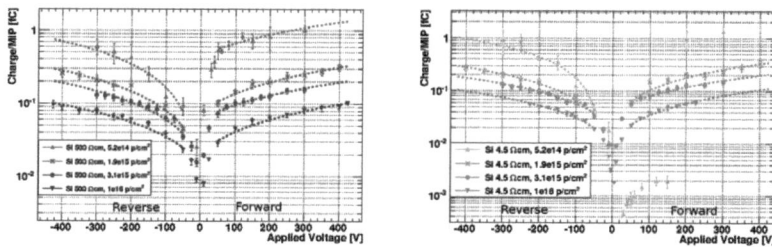

Figure 6.28: Voltage scan for 500 Ωcm silicon detector on the left and for 4.5 Ωcm silicon detector on the right.

3. Decrease of the slope, because even high voltages did not allow a full charge collection efficiency

Non-irradiated single crystal CVD diamond detectors have a full charge collection above 100 V, corresponding to a small voltage scan slope of almost 0. This is also the case for non-irradiated silicon detectors, once full depletion voltage is reached. The peak is therefore also estimated to have occurred for silicon material at lower fluencies. The peak indicates that up to a certain fluence it is still possible to obtain a full charge collection by the application of higher voltages, but this ability is reduced through more irradiation and thus more radiation defects. These observations are coherent with the variation of the fitting parameters of the degradation curves with respect to applied voltages. An analysis of the k parameters for the silicon detectors degradation curves, suggests that the peak of the slope in figure 6.29 for silicon material should be around a fluence of $3 \cdot 10^{14}$ protons/cm^2.

The fact that sCVD had the largest slope at high fluences, means that it had the largest potential to collect further charges through the application of higher voltages and that its saturation voltage was not reached with the applied voltages.

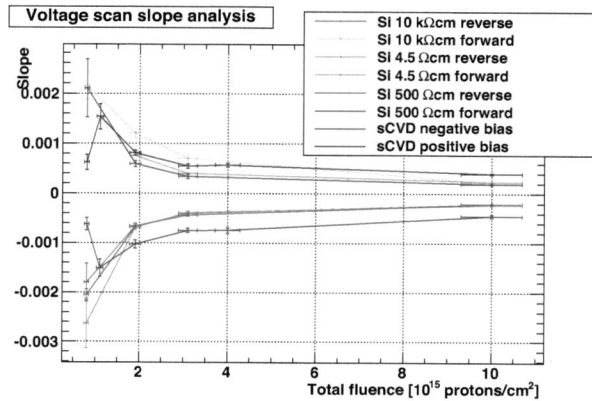

Figure 6.29: Slopes from the voltage scan fits plotted as a function of the fluence.

Figures 6.30 and 6.31 show how the parameters vary with the applied voltage. The

k parameter had the general tendency to decrease with increased voltage. This was especially pronounced for the silicon detectors under reverse bias. To notice from the s plot is the significant increase with higher voltages for the 4.5 Ωcm silicon detector sample under reverse bias modus.

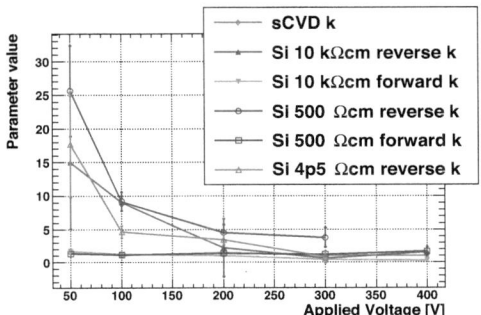

Figure 6.30: Dependence of the fitting parameter k on the applied voltage.

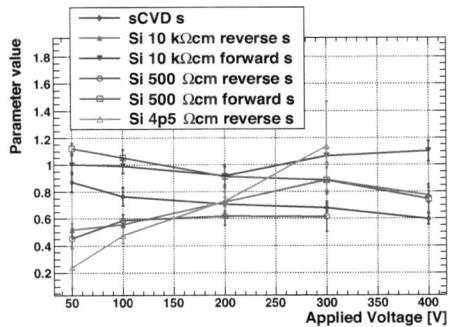

Figure 6.31: Dependence of the fitting parameter s on the applied voltage.

Linearity of the detector response

At a fluence of $6.8 \cdot 10^{15}$ protons/cm^2, the beam intensity was reduced by a factor of 7 for testing purposes and to lower the influence of additional radiation damage during the 80 K warm-up measurements described below. Figure 6.32 shows the signal dependence on the number of protons before the warm-up. The lines going through the origin, show that at cryogenic temperatures no significant non-linearities are observed due to high charge density. For silicon detectors at room temperature the low initial charge yield was explained by a high charge carrier density effect (plasma effect). This is not the case in liquid helium and was confirmed by this linearity check. The most probable reason is that the silicon material charge mobilities are higher by about 3 orders of magnitude at liquid helium temperatures (see section 4.5).

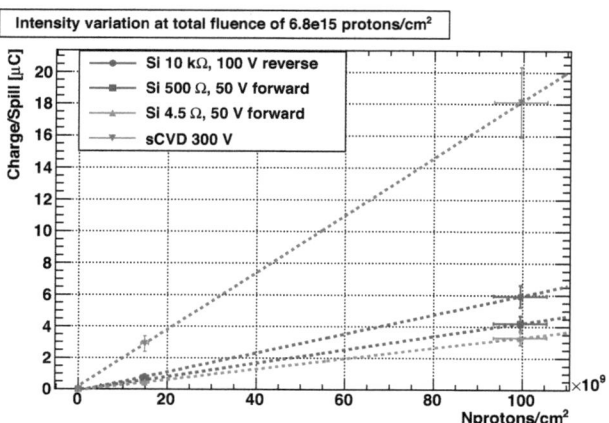

Figure 6.32: Detectors collected charge under beam intensity variation at 4.2 K and a total fluence of $6.8 \cdot 10^{15}$ protons/cm^2.

Warm-up cycles

A maximum warm-up to 80 K is expected in the LHC magnets during the LHC winter shutdowns over the Christmas period. This temperature is chosen because the mechanical stress due to thermal expansion on the magnet parts from 1.9 K to 80 K are still at a minimum.

Therefore a first warm-up cycle to 80 K at a fluence of $6.8 \cdot 10^{15}$ protons/cm^2 was performed. The measured changes of the detector properties up to 80 K were not significant. The leakage current and the collected charge were at the same level as at 1.9 K. No significant annealing of the detectors can hence be expect for the CryoBLM application in the LHC during the winter shut downs. After the warm-up to 80 K and the performance of all planed measurements, the cryostat was filled again with liquid helium and the beam intensity was increased back to the nominal value.

During the final warm-up to room temperature the 10 kΩcm silicon detector signal recovered by a factor of 7.5 ± 1.2, compared to the one in liquid helium as shown in figure 6.33. Only above 135 K a significant signal recovery is noticeable for the silicon detectors. Part of the induced signal degradation is hence reversible in silicon material, but the collected charge does not reach the same value as at low dose. Some of the damage is therefore permanent, especially for the diamond detectors.

Diamond material is less temperature dependent than silicon material. The diamond detector signal did not recover during the warm-up cycles. This was already observed by a group performing diamond material 2 MeV/c electron irradiation at 90 K with the purpose of defect analysis [103]. Particle detection measurements were not in the scope of their investigations. They did not observe any defect modifications or annealing effects when warming up the irradiated diamond material from 90 K to room temperature.

The leakage current increase with temperature for the silicon detector under 100 V applied voltage is shown in figure 6.34.

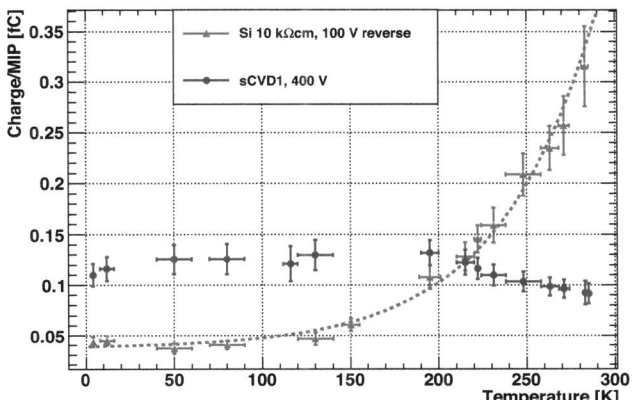

Figure 6.33: Collected charge from the silicon and diamond detectors during the warm-up of the cryostat. The signal recovery started with the re-apparition of a measurable leakage current, indicating a link between the two effects.

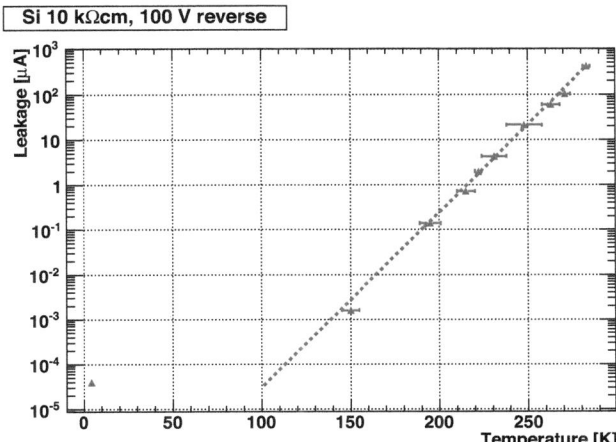

Figure 6.34: Silicon leakage current during warm up from liquid helium temperatures to room temperature. The measurement is biased by the fact that irradiation took place during the warm up, therefore the leakage current increase is not only due to the temperature increase, but also due to increased irradiation (the fluence increase is of 7 % from the beginning of the warm-up to the end). The warm-up to RT lasted 50 hours during which the beam was kept on to measure the temperature dependence of the detector signals.

Degradation curve comparison

The degradation curve comparison with other irradiation experiments on diamond detectors is shown in figure 6.35. As diamond detector comparison data, the curves from Moritz Guthoff using pCVDs and sCVDs in the CMS experiment [104] and from RD42 have been used. The diamond detectors in the CMS experiment were also operated in DC mode, while for the RD42 charge collection measurements a β-source and a scintillator as external triggers were used [95]. The initial collected charge $Q(0)$ at zero fluence has been normalised to one for all detectors, in order to compare the trends with increased radiation. The differences are significant, especially between the measured performance of RD42 compared to the others.

The comparison between sCVD and pCVD at room temperature is as expected, because pCVD is like a pre-damaged sCVD, with smaller initial collected charge, but also smaller signal reduction with higher fluences.

The observation that sCVD had a better performance in liquid helium was not expected. On the contrary it was feared that due to increased polarisation effects the sCVD would not be able to collect any charges. One possible explanation for this observation is that the effect of polarisation was counterbalanced by the fact that the traps were filled and in liquid helium no charge detrapping, even of shallow levels, was occurring. The traps were therefore neutralised and could not trap further charges.

This is of advantage, as it means that the sCVD in cold has the same initial charge collection as at room temperature, but its signal decrease with irradiation is rather comparable to the one of pCVD at room temperature.

In the RD42 collaboration around H. Kagan, one tested diamond detector had poor radiation hardness. The problem could be solved by removing the contacts and 5 μm of diamond material by Reactive Ion Etching (RIE) on each side of the detector. Subsequently the detector was re-metallised and full signal recovery could be observed. This leads to the preliminary conclusion that the observed signal degradation, together with the introduction of the additional parameter s, might be due to surface effects and not due to radiation damage in the detector bulk. The challenge for the CryoBLM application is to find a reliable way of detector characterisation, as leakage current, TCT and beta-source measurements were according to the expectations for the detectors with poor radiation hardness properties. This means that possible surface issues only appear during the irradiation, which is problematic as in the CryoBLM project no detector exchange or mechanical manipulation is foreseen for the moment. A potential to systematically avoid the surface issue could be by optimised doping of the contacts, as already started for the silicon detectors. To better understand these effects, measurements on the irradiated detectors are planed and further investigation on diamond detectors with boron doped contacts is ongoing.

The degradation curve comparison with other irradiation experiments on silicon detectors is shown in figure 6.36. The data for silicon irradiation at room temperature is taken from [95, 105]. The shown model uses the Hecht equations based on charge mobility and trapping probability.

6.5.3 TCT observations

Along with the measurements of DC characteristics, current pulse response was recorded using TCT. The observed signal degradation measured with TCT confirmed the rather fast signal decrease as shown in DC-operation. The preliminary results also showed that space charge sign inversion occurred at relatively low fluence as it is the case for RT irradiation.

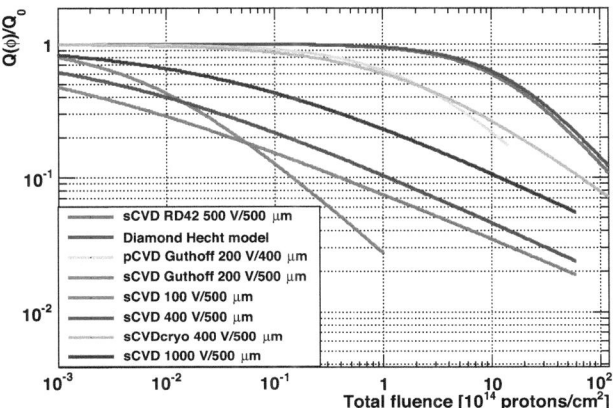

Figure 6.35: Comparison of the diamond detector degradation curves between the room temperature irradiation, the liquid helium irradiation, the degradation model and measurements performed by other groups. At a zero fluence all normalised charges equal 1.

6.6 Summary

An experimental procedure was developed to perform the first long term radiation tests of solid-state detectors in liquid helium at the CERN irradiation facility. The leakage current for silicon detectors at cryogenic temperatures decreased to 100 pA at 400 V, even under forward bias. The radiation effect on the detector sensitivity for silicon and single crystal CVD diamond detectors was measured and a model for the signal degradation using the results from Chapter 4 was derived and applied to the irradiation data. The expected reduction in signal over 20 years (2 MGy) of LHC operation was a factor of 52 ± 11 for the silicon device at 300 V reverse and a factor of 14 ± 3 for the diamond detector at 400 V. For the silicon detector under 300 V forward bias, the charge yield was superior to the normal operation mode and the signal reduction is 25 ± 5. Diamond detectors had a higher signal than silicon sensors after an irradiation of 0.9 MGy. This level of radiation hardness of the diamond and forward silicon detectors are acceptable for the CryoBLM application, but a calibration has to be foreseen, as the beam abort threshold setting of the system has to be adapted to the signal reduction of the detector.

During the warm-up to RT, the silicon detectors recovered their signal by a factor of 7.5 ± 1.2, while the diamond detectors did not recover with increased temperature.

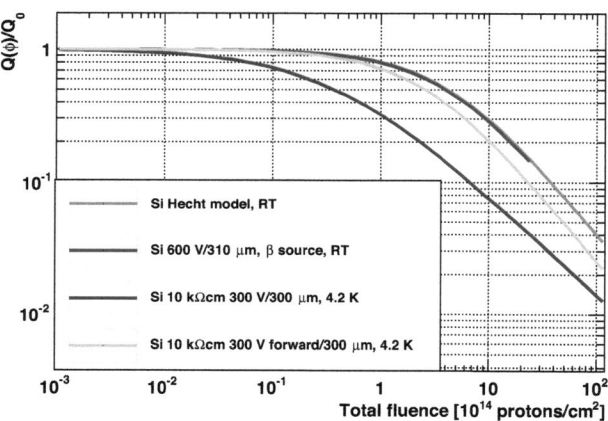

Figure 6.36: Comparison of the silicon detector degradation curves between the liquid helium irradiation, the degradation model and the measurements performed by other groups.

CHAPTER 7

Liquid helium chamber measurements

An introduction to the charge carrier properties in liquid helium is given in section 4.5. The charge carriers in liquid helium build up special structures, electrons form "e-bubbles" and ions become "snowballs". These structures contain up to several hundred of helium atoms, which slow down the drift of the charge carriers. In this Chapter, the measurements done with the liquid helium chamber prototypes in beam are summarised. The data analysis and the main results are discussed, allowing to analyse the advantages and disadvantages of this detector type.

7.1 Experimental setup

7.1.1 Beam lines

Beam tests have been performed in two different beam lines at CERN. The beam lines have been introduced already in chapter 5 for the low intensity beam line and in section 6.4 for the high intensity beam line used for irradiation experiments.

7.1.2 Signal readout and data acquisition

The number of protons was measured with a scintillator for the low intensity beam tests and with a Secondary Emission Chamber (SEC) [99] during the high intensity beam tests.

This measurement procedure allowed to measure the low currents from the low intensity beam, but could not be used for the investigation of the liquid helium chamber charge collection time. This was accomplished with a direct connection of the signal side of the detectors to a 1 GHz oscilloscope from LeCroy.

In the low intensity beam the cable lengths from detector to the measurement device were below 15 m, while for the high intensity beam the length was of 50 m using standard CB-50 cables with BNC connectors.

7.1.3 Detector prototypes

The liquid helium chambers were designed based on the existing BLM ionisation chambers, for which the detection medium is nitrogen gas. The size of the first prototype was adapted in order to get a similar total charge collection from the liquid helium as from the nitrogen gas in the larger BLM chamber.

The chambers consist of parallel metallic plates connected with metallic stabilisation rods. The large BLM nitrogen ionisation chamber has six stabilisation rods, while the liquid helium chamber only has four rods. These rods guarantee the mechanical stabilisation of the plates and enable the electric connectivity of each plate. The plates and the rods of the large chamber were made out of stainless steel. The metallic spacers between the plates were made of nickel plated brass, while the used isolating material was nylon 6-6. For the small ionisation chamber, the metallic material was aluminum, while the isolating parts were made of ceramic.

The two tested detector prototypes are shown in figure 7.1.

Figure 7.1: Ionisation chambers based on liquid helium as detection medium. The larger one is the first prototype. It is made out of stainless steel and the distance between the plates is of 3 mm. It allowed the prove of principle. The smaller ionisation chamber is made out of aluminum and the distance between the plates is of 1 mm.

The reasons for building the second liquid helium chamber prototype were the following:

- charge drift time reduction by decreasing the distance between the plates,
- reduction of the activated material by reducing the thickness of the plates and by using aluminum instead of stainless steel,
- reduction of the size of the chamber due to limited available space and
- installation of a metallic housing for mechanical protection and additional electric shielding.

The active length of the larger helium chamber is of 3.9 cm and its active area of 15.9 cm^2. The active length of the smaller helium chamber is of 2.45 cm and its active area of 12.6 cm^2. The total active cross section of both chambers were always larger than the area occupied by the beams inside the chambers.

The capacitance C of the larger helium chamber was measured at room temperature in air to be of 107 pF, while the smaller one was of 227 pF. The difference between the two is due to their different areas A and distances between the plates d, as shown in the following equation:
$$C = \varepsilon_r \varepsilon_0 \frac{A}{d} \tag{7.1}$$
where ε_r is the relative permittivity of the material between the plates and ε_0 is the relative permittivity of vacuum.

From 1.75 K to 4.2 K the values for the dielectric constant (equivalent to the static relative permittivity) of liquid helium are between 1.048 and 1.056 [106]. The detector capacitance is therefore about 5 % higher in liquid helium than in air. This variation in capacitance has no effect on the charge collection efficiency and on the timing properties of the detector.

7.2 Results

Several measurements could be done, allowing the prove of principle and allowing the measurement of basic properties of the liquid helium chamber prototypes.

7.2.1 Collected charge per MIP

The results were obtained using the measurements with the DC readout, as already discussed in section 6.4.

In section 4.5.5 the charge was estimated to be of 1.14 fC/cm per MIP. Figure 7.3 shows the results from a voltage scan on the liquid helium chamber in beam. The charge collected per MIP at 1.8 K with an applied electric field of 200 V/mm was of 0.115 $\pm$ 0.01 fC/cm, which is a factor 10 below the expectations. The fit shows that the saturation was not yet reached, which is one of the explanations for the difference between measured and estimated charge collection. One also notices the difference in charge collection between 4.2 K and 1.8 K. The reason for this observation is not known. Figure 7.2 shows the measurements taken at different beam intensities. The linear fit yields a charge per MIP of 0.106 $\pm$ 0.006 fC/cm at a temperature of 1.8 K and an electric field of 200 V/mm. This value is in agreement with the collected charge measured during the voltage scan.

7.2.2 Time response

The spill duration in the available beams was of 400 to 450 ms, without any sharp edges in the time structure. Sharp edges would allow to obtain the exact delay of the liquid helium chamber compared to the silicon detector. Ideal would be a beam with a short delta peak of protons, allowing to see the drift of the ionised charges through the liquid, without any effects from the spill shape. This was not possible to achieve with the PS beam and it was not possible to transport the cryogenic equipment with the required amount of liquid helium to other facilities.

In figure 7.4 one of the sharpest falling edges of about 1 ms from spill is shown. A linear fit to the falling edge has been applied, in order to distinguish between the signal of the protons from the spill and the signal due to the drift of the charges. This very direct method is not useful in this case, because the timing is dominated by the spill shape.

During the first low intensity beam tests, the timing properties of the liquid helium chamber were investigated. In the low intensity beam test an average of 88 protons arrived within the time of interest of 100 µs. This corresponds to a measured current of only 0.14 pA. Three

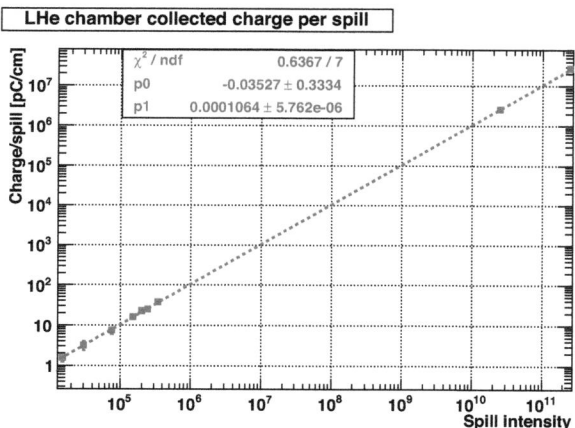

Figure 7.2: Linearity of detector response as a function of the spill intensity.

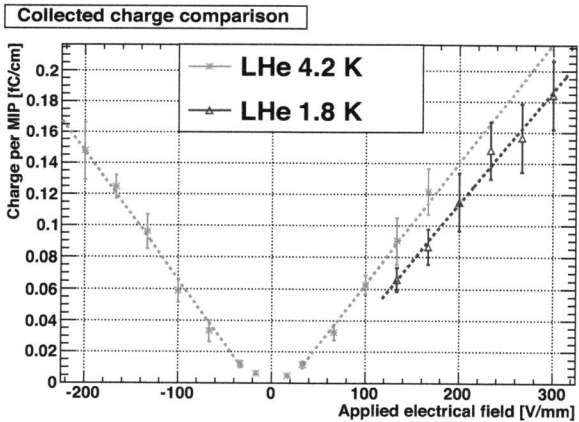

Figure 7.3: The plot shows the measured collected charge per MIP as a function of the applied electric field in liquid helium and superfluid helium. Within the measurement range, saturation is not reached.

different amplifiers were used, but the very low signal to noise ratio (SNR) together with the long spill duration made any conclusions impossible. An additional effort was put into the offline analysis through the use of Fourier transformations, moving average and moving median techniques with the aim to improve the SNR. It was not possible to obtain a balance between noise cancellation to optimise the SNR and keeping the spill time structure information down to 100 μs. A visual summary of the signal measurements with low intensity beam is depicted in figure 7.5, where the untreated data is shown with its large noise component together with the treated data in the right plot.

Only with the high intensity beam from the T7 beam line, conclusions about the timing

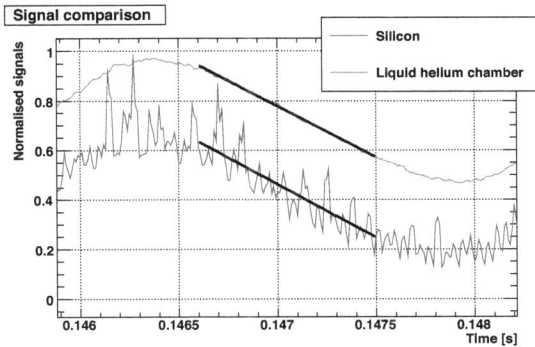

Figure 7.4: Example of signals from the liquid helium chamber and the silicon detector during a small part of the total spill. The signal drop was one of the most pronounced ones that could be found. The decrease was linearly fitted (shown with black lines) with the intention to obtain a significant difference due to the slower charge carrier drift in liquid helium. The method is not successful, because the reduction of number of protons in the spill is not fast enough.

structure of the charges in liquid helium could be drawn. While the SNR in the T9 beam line was below 1, the average SNR in the T7 beam line was above 30 for the LHe chamber, due to large number of protons within a T7 spill.

The sampling rate of the 1 GHz LeCroy oscilloscope was set to 100 kHz, which allows conclusions about the liquid helium charge collection time.

The signal in a silicon detector at 1.9 K from a minimum ionising particle has a FWHM of 2.5 ± 0.7 ns [96]. It can therefore be considered as instantaneous compared to the expected charge collection time in liquid helium.

The silicon sensor used for comparison was a 10 kΩcm p^+-n-n^+ diode with a reverse applied voltage of 100 V. The thickness of the detector was of 300 µm and its active area was of 23 mm^2.

It should further be mentioned that both detectors were within the centre of the Gaussian beam and the liquid helium chamber in addition measured the transversal tails of the beam too, due to the difference in size between the detectors. This should not significantly influence the time structure analysis, which is further confirmed by the quality of the obtained results.

Figure 7.6 shows the time distribution of the spill measured with the silicon detector and the small liquid helium chamber. The same spill shape is shown in figure 7.7 with the application of a zoom to better visualise the structure of the spill.

A time shift between the maximum and minimum of the silicon detector and the liquid helium chamber could be due to different cable lengths from the detector to the oscilloscope or an actual physical distance between the two detectors, where the particles would arrive at different times at the detectors. The physical distance between the middle of the silicon detector and the middle of the liquid helium chamber was below 6 cm, corresponding to about 0.3 ns delay. The difference between the total cable lengths from the detectors to the measuring instrument was below 2 m, corresponding to a 10 ns delay. Both delays were neglected with respect to the estimated time constants of the liquid helium chamber.

The signal expected from a MIP in a detector is a triangle (as shown in figure 7.8), having

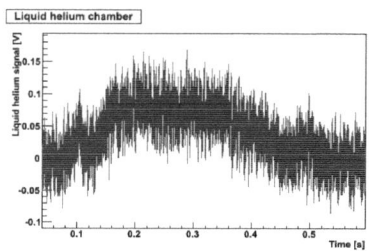 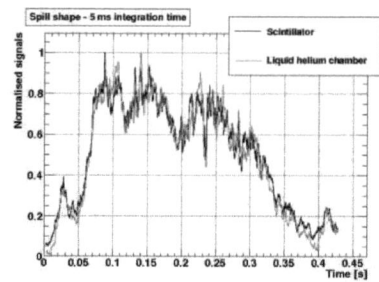

Figure 7.5: Example of the time distribution of the spill from the T9 beam line. On the left side the liquid helium chamber signal at 1.9 K and at an electric field of 330 V/mm is shown. The low SNR is visible, especially looking at the beginning and the end of the plot, where only noise and no particle signals are present. The right panel shows the treated and normalised signals of the liquid helium chamber and a scintillator placed before the cryostat for the same spill. The shapes on the right were obtained with 5 ms integration times, and therefore deteriorating the time structure information. The intensity of the spill was of $3.1 \cdot 10^5$ protons/cm^2 during 430 ms.

a sharp rise time (limited by electronics for semiconducting detectors) and a linear decrease until all charges are collected. This can be theoretically shown using Ramos theorem [107]. The sharp rise time is due to quasi instantaneous generation of charges by ionisation of the atoms along the path of the particle. The linear decrease of the signal is due to the drift of the charges, where the ones generated right next to the electrode of opposite polarity have a shorter drift than the ones generated next to the electrode of same polarity. Once the charges are collected at the electrode, their contribution to the signal disappears leading to the linear signal decrease with time. If the charges have different drift times, these can be distinguished in the shape of the signal response, as measured during the signal particle detection measurements in section 5.4.3.

In order to compare the signal from the silicon diode with the signal from the liquid helium chamber, the strategy was to offline manipulate the fast silicon detector spill shape in such a way, that it would fit the LHe chamber spill shape. Therefore an artificial pulse shaping was introduced, corresponding to a drift deceleration for the charge carriers in the silicon material.

The signal at a moment t is a combination of signals from charges generated by particles traversing the detector at the time t and particles having traversed at earlier times $t - t'$, where t' depends on the charge velocity v and the distance to the electrodes of the detector. The maximum of t' taken into account corresponds to the charge drift time t_{drift} from one detector electrode to the next with $t_{drift} = \frac{d}{v}$, where d is the distance between the plates. The charge velocity is the product of the charge mobility μ and the applied electric field E: $v = \mu E$. While this charge drift time is very short for silicon material, it is long inside liquid helium. In order to artificially decelerate the drift of the charge carriers in silicon material, one therefore has to take into account earlier signals in the offline analysis.

The proposed method to simulate the artificially lowered drift velocity t_{drift} of the charge was the following:

$$S_{slowSi}(t) = \frac{\int_{t-t_{drift}}^{t} w(t') S_{Si}(t') dt'}{\int_{t-t_{drift}}^{t} w(t') dt'} \qquad (7.2)$$

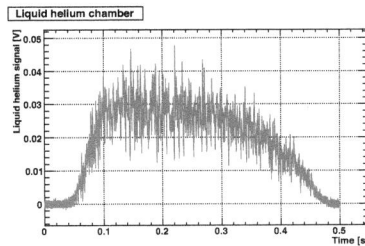 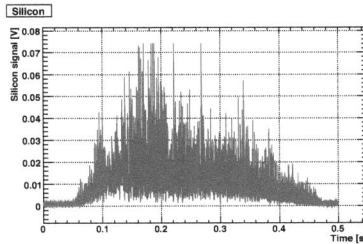

Figure 7.6: Shape of an example spill. On the left is the complete spill measured with the liquid helium chamber and on the right is the same spill measured with the silicon detector. The shapes look very different, which was due to two main reasons. First the fact that the charges drift slower through the liquid helium chamber and second the difference in size of the detectors.

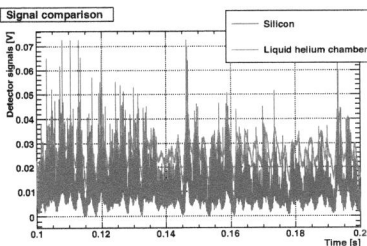 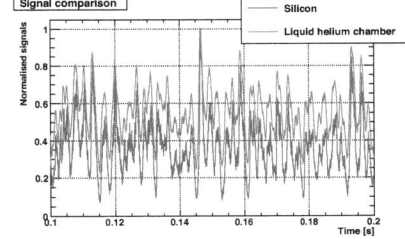

Figure 7.7: Zoom into the spill from figure 7.6 and direct comparison of the detector signals. On the left is the normalised original signal, where the spill fine structure is visible with the silicon diode. On the right is the transformation of the silicon detector signal with an artificial charge carrier drift time of 180 μs.

with $w(t')$ being the weight, with which the earlier signals were taken into account:

$$w(t') = \frac{t - t'}{t_{drift}} \qquad (7.3)$$

A linear weighting factor was chosen, as a perfect triangular shape of the signal was assumed. This means the earlier the signal was, the less weight it has and the less it is taken into account. This has been depicted in the dotted green line of the simplified drawing in figure 7.8.

After this procedure of artificial slowing down of the silicon detector charge carriers, both signals - the slowed silicon signal and the liquid helium chamber signal - were normalised so that their maximum signal equals one. It should be mentioned that the further described method is independent of the normalisation. The next step was to test for which drift velocity the signal from the silicon detector best agrees with the one of the liquid helium chamber. In order to quantify to what degree the signals agree, the correlation method was chosen. The correlation factor $corr_{xy}$ is defined as:

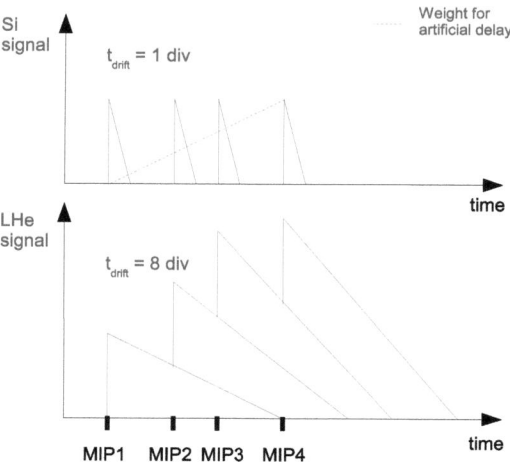

Figure 7.8: Simplified picture of the signals in the detectors from 4 minimum ionising particles (MIP1 to MIP4), arriving at different times. For the silicon detector the signal amplitude at the arrival time of MIP4 is only due to MIP4, while for the LHe chamber, the signal amplitude at the same time is due to the earlier MIPs as well.

$$corr_{xy} = \frac{\sum\limits_{i=1}^{n}(x_i - \bar{x})(y_i - \bar{y})}{\sqrt{\sum\limits_{i=1}^{n}(x_i - \bar{x})^2 \sum\limits_{i=1}^{n}(y_i - \bar{y})^2}} \qquad (7.4)$$

where x stands for the slowed silicon signal and y for the liquid helium chamber signal. n is the total number of measurement points. The corresponding error of the correlation factor is:

$$s_{corr} = \frac{1 - (corr_{xy})^2}{\sqrt{n-1}}; \qquad (7.5)$$

The closer the correlation factor $corr$ is to 1, the better the two signals correspond to each other. These calculations were repeated for different artificial charge carrier drift manipulations in order to find the drift time that best describes the liquid helium chamber signal.

Figure 7.9 shows the chamber signal plotted versus the silicon signal at same measurement times. The more incoherently spread out the points are, the less the signals correspond to each other. With increasing charge drift time a higher correlation can be visually noticed, while at some point the slowed charge drift becomes too large and the points start to be more scattered again.

The result of the correlation calculation for different values of the artificial slowed charge drift is shown in figure 7.10. The maximum correlation was at 180 μs for an electric field of 400 V/mm and at 280 μs for 200 V/mm. This is consistent with the fact that for lower voltages the charges drift more slowly. The expected charge collection time from mobility calculations in section 4.5.6 was of 290 μs at 400 V/mm and of 550 μs at 200 V/mm. The obtained timing from the measurements were below these calculated values as the calculations were

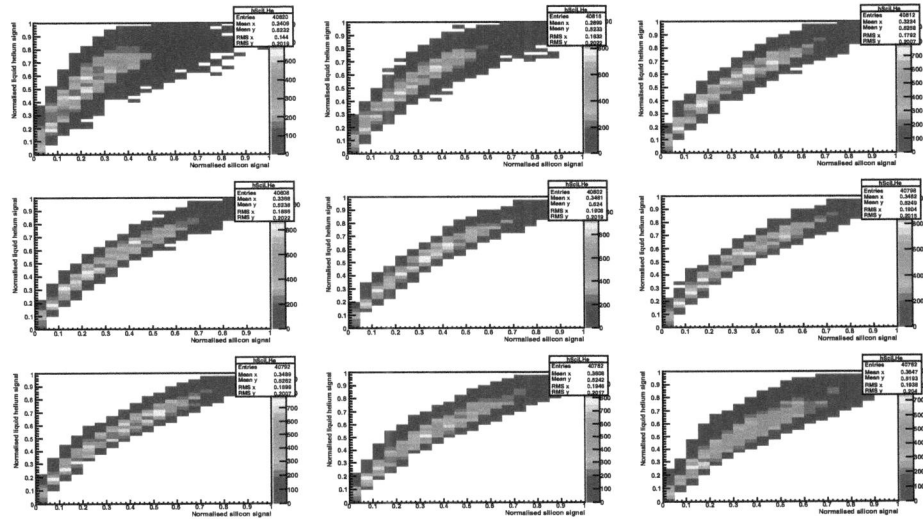

Figure 7.9: Visualisation of the analysis procedure. The normalised signals from the liquid helium chamber (Y-axis) and the silicon detector (X-axis) with same time stamps are plotted against each other for different integration times. The artificial charge drift time goes from 20 μs in the upper left plot to 600 μs in the lower right plot.

done for charges drifting over the whole distance between the plates, therefore corresponding to a full charge collection time. This is a timing overestimation, as the charges are generated evenly between the plates for MIPs, resulting on average in half of the maximum drift time. This is rather well born out by the measurements. In addition the maxima from the method of the slowed charged drift in silicon material (figure 7.10) are not very pronounced which is due to the nature of the experiment with 400 ms spill duration and due to the fact that ions and electrons have only slightly different drift times as shown in figure 4.21. Therefore one would ideally expect two distinct maxima, which in this method is seen as one flat maximum, corresponding to a mean charge drift time. The optimal slowed silicon signal is shown in figure 7.7, where the similarity to the liquid helium chamber is much more pronounced than for the original signal.

7.3 Summary

The measured charge collection at 1.8 K with an applied electric field of 200 V/cm was of 0.115 $\pm$ 0.010 fC/cm per MIP. Due to this small generated charge per MIP and the slow drift of the charge carriers in liquid helium, the timing properties were difficult to measure, but the manipulation of the signal shapes allow to extract a measurement of the drift velocity, which was found to be close to the theoretical value. First results in beam showed that the current design of the liquid helium chamber enables protection from steady state losses (happening in the order of seconds) and any losses that are slower than about 180 μs.

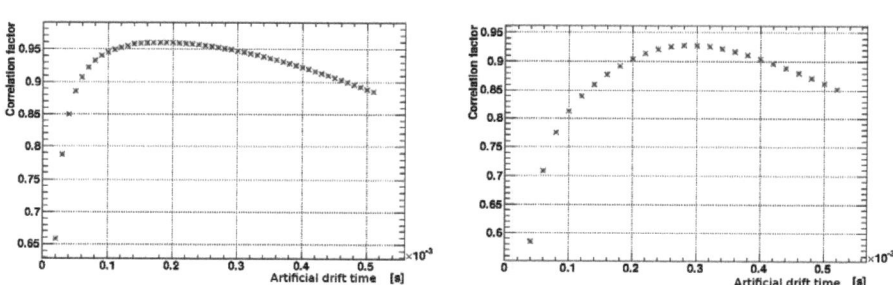

Figure 7.10 The calculated correlation factors plotted versus the artificial charge drift times of the silicon material. A maximum is reached for a charge drift of 180 μs at an electric field of 400 V/mm on the left side, while the maximum is at 280 μs at a field of 200 V/mm seen on the right side. That no sharp maxima are obtained is due to the fact that there are two maxima close to each other, one from ions and one from electrons, that can not be resolved with this experiment and this method.

CHAPTER 8

Conclusion and outlook

The goal of this thesis was the investigation of detectors for the use as cryogenic beam loss monitors inside the LHC magnets. The research started by an evaluation of the promising detector technologies and continued through the design of experiments at room temperature and under cryogenic conditions to allow a better understanding of the detector properties and their suitability as future CryoBLM. The candidates under investigation in this work were diamond and silicon detectors and an ionisation chamber, using the liquid helium itself as particle detection medium. The reasons for their selection were summarised in Chapter 3.

The ability of the selected detectors to measure radiation depends on charge transport phenomena within the active medium. Experiments were therefore performed in the laboratory with a laser and an alpha source, providing a detailed picture of the charge carrier properties in the solid-state detectors down to LHe temperatures. The temperature dependence of the drift velocity and of the mobility of the charge carriers was measured. These measurements and results were presented in Chapter 4.

In Chapter 5 the measurements of single particles with solid-state detectors at cryogenic temperatures were outlined. For MIPs the pulse FWHM at liquid helium temperatures is of 2.5 ± 0.7 ns for silicon detectors and of 3.6 ± 0.8 ns for diamond detectors. This confirms the viability of a cryogenic detection system allowing bunch by bunch resolution for the LHC.

The radiation hardness of the solid-state detectors was measured in DC mode during irradiation measurements at room temperature and at liquid helium temperatures. The results of the irradiations were summarised in Chapter 6. The expected reduction in signal over 20 years (2 MGy) of LHC operation is a factor of 25 ± 5 for the silicon device at 300 V and a factor of 14 ± 3 for the diamond detector at 400 V. The silicon detector's forward bias characteristics were measured in detail for the first time at liquid helium temperatures with laser, with beam and under irradiation. The forward current at 1.9 K was of 98 µA at 200 V and decreased below 100 pA when the diodes were irradiated. Its initial resistance to irradiation was better than the one detector under reverse bias.

Two liquid helium chamber prototypes were constructed allowing the proof of principle and the characterisation during several beam tests, as presented in Chapter 7. With the current design of the liquid helium chamber a successful protection from losses with a time constant above 180 µs is ensured.

With silicon and diamond sensors a fast detection system can be designed allowing bunch by bunch resolution. The LHe chamber on the other hand is an elegant solution due to its insensitivity to radiation damage. The results hence show the advantage of combining the two approaches, by using solid-state detectors for a fast protection system, while the LHe chamber can be used in parallel for calibration and for the protection from steady state losses.

During 2012 the first installation of two silicon and two diamond detectors on the cold mass of a LHC magnet was performed and can be seen in figure 8.1. Further installations of CryoBLMs in the tunnel are planned during the Long Shut-down 1 (LS1).

These first LHC CryoBLM detectors will allow testing of:

- the detectors performance,
- their long term stability and
- their radiation hardness for actual LHC particle showers and particle rates.

In addition their symmetric positions close to the beam line will give an unprecedented insight into beam losses in the LHC.

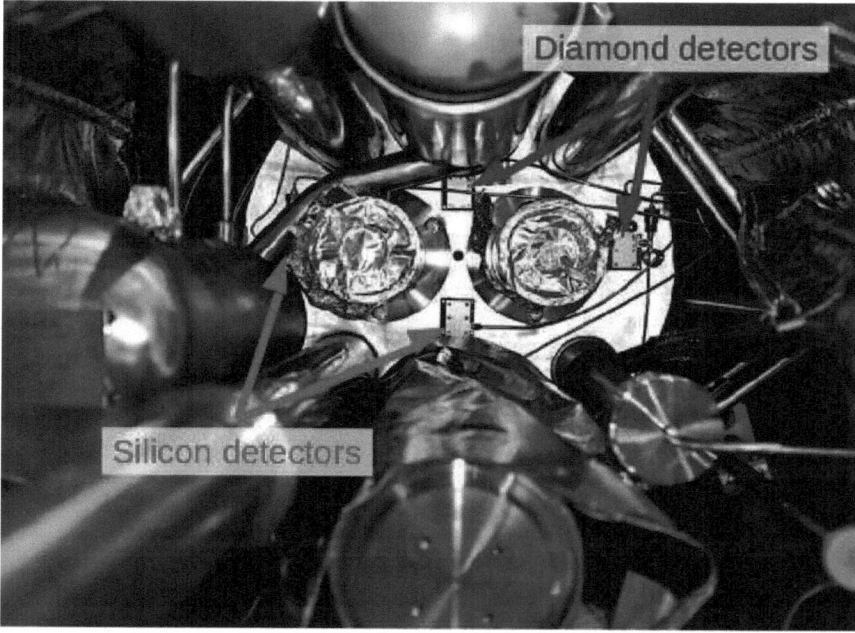

Figure 8.1: Installation of two silicon detectors and two diamond detectors on the cold mass of the magnet Q7R3. A good metallic contact is given between the detectors and the cold mass of the magnet to guarantee that the detectors are operated at the temperature of 1.9 K.

Bibliography

[1] Atlas Collaboration: *Observation of a new particle in the search for the Standard Model Higgs boson with the ATLAS detector at the LHC*, Physics Letters B, Volume 716, Issue 1, 17 September 2012, Pages 1-29.

[2] CMS Collaboration: *Observation of a new boson at a mass of 125 GeV with the CMS experiment at the LHC*, Physics Letters B, Volume 716, Issue 1, 17 September 2012, Pages 30-61.

[3] *LHC Design Report, Volume 1*

[4] T. Baer et al: *UFOs in the LHC*, Proceedings of IPAC 2011, ISBN 978-92-9083-366-6, p. 1347-1349, TUPC137, San Sebastián, Spain, September 2011.

[5] T. Baer et al: *UFOs in the LHC after LS1*, Proceedings of Chamonix workshop on LHC performance, Chamonix, France, p. 294-298, Feb. 2012.

[6] D. Bocian, B. Dehning, A. Siemko: *Modeling of Quench Limit for Steady State Heat Deposits in LHC Magnets*, IEEE Transactions on Applied Superconductivity, Volume 18, Issue 3, p. 2446 - 2449, 2009.

[7] J. B. Jeanneret, D. Leroy, L. Oberli, T. Trenkler: *Quench levels and transient beam losses in LHC magnets*, LHC Project Report 44, 1996.

[8] M. Sapinski et al: *Quench Limits*, Proceedings of Chamonix workshop on LHC performance, Chamonix, France, p. 200-211, Feb. 2012.

[9] A. Priebe: *Quench behaviour modelling of the LHC superconducting magnet*, PhD thesis in preparation.

[10] C. Kurfuerst et al: *Particle Shower Simulations and Loss Measurements in the LHC Magnet Interconnection Regions*, Proceedings of IPAC 2010, ISBN 978-92-9083-352-9, p. 2857-2859, WEPEB070, Kyoto, Japan, 2010.

[11] V. Kain: *Machine Protection and Beam Quality during the LHC Injection Process*, PhD thesis, University of Vienna, 2005.

[12] W. Demtröder: *Experimentalphysik 1.*, Fifth Edition, ISBN: 978-3-540-79294-9, Springer, 2008.

[13] D. H. Wilkinson: *Ionisation chambers and counters*, Cambridge Monographs on Physics, Cambridge, 1950.

[14] G. F. Knoll: *Radiation Detection and Measurement*, Third Edition, ISBN: 0-471-07338-5, John Wiley & Sons, 2000.

[15] M. Stockner: *Beam Loss Calibration Studies for High Energy Proton Accelerators*, PhD thesis, TU Vienna, 2007.

[16] D. Kramer: *Design and implementation of a detector for high flux mixed radiation fields*, PhD thesis, 2008.

[17] C. Kurfuerst: *Quench Protection of the LHC Quadrupole Magnets*, Diploma thesis, TU Vienna, 2010.

[18] M. Sapinski et al: *Simulation of Beam Loss in LHC MB Magnet and Quench threshold test*, LHC Project Note 422, CERN, Geneva, Switzerland, 2009.

[19] M. Sapinski et al: *Estimation of thresholds for the signals fo the BLMs around the LHC final focusing triplet magnets*, Proceedings of IPAC 2012, ISBN 978-3-95450-115-1, p. 4053-4055, New Orleans, USA, May, 2012.

[20] N. V. Mokhov et al: *Protecting LHC IP1/IP5 Components Against Radiation Resulting from Colliding Beam Interactions*, LHC Project Report 633, CERN, Geneva, Switzerland, 2003.

[21] L. Sarchiapone et al: *FLUKA Monte Carlo simulations and benchmark measurements for the LHC beam loss monitors*, Nucl. Instr. and Meth. A, Vol 581, Issues 1-2, p. 511-516.

[22] S. Roesler et al: *The Monte Carlo Event Generator DPMJET-III*, Proceedings of the Monte Carlo 2000 Conference, Lisbon, October 23-26 2000, Springer Verlag Berlin (2001) p. 1033-1038.

[23] M. Sapinski: *First look at the quench test results*, LHC Machine Committee, CERN, 27.03.2013.

[24] Ch. Erss et al: *Cryogenic Particle Detection*, Topics in Applied Physics, Volume 99, Springer, ISBN 3-540-20113-0, 2005.

[25] S. W. Van Sciver: *Helium Cryogenics*, The International Cryogenics Monograph Series, Plenum Press, 1986.

[26] http://www.detectors.saint-gobain.com, 17.04.2013, 16:00.

[27] J. Bosser et al: *Preliminary Measurements on Microcalorimeters foreseen to be used as Beam Loss Monitors*, LHC Project Note 71, CERN, Geneva, Switzerland, 1996.

[28] D. N. McKinsey et al: *Detecting ionizing radiation in liquid helium using wavelength shifting light collection*, Nuclear Instruments and Methods in Physics Research A 516, p. 475-485, 2004.

[29] http://web.mit.edu/figueroagroup/ucal/ucal_tes/index.html, 14.10.2011, 15:00.

[30] T. Niinikoski and F. Udo: *"Frozen Spin" Polarized Target*, Nucl. Instr. and Methods I34, p. 219-233, 1976.

[31] G. Lutz: *Semiconductor Radiation Detectors*, Springer, ISBN 3-540-64859-3, 1999.

[32] M. Dragicevic: *The New Silicon Strip Detectors for the CMS Tracker Upgrade*, PhD thesis, TU Vienna, 2010.

[33] H. Spieler: *Semiconductor Detector Systems*, Oxford Science Publications, ISBN 978-0-19-852784, 2005.

[34] K. Wittenburg: *The PIN-diode Beam Loss Monitor System at HERA*, DESY-HERA-00-03, Jun 2000.

[35] RD39 Collaboration: *RD 39 status report 2006*, CERN-LHCC-2006-034, 2006.

[36] RD39 Collaboration: *RD 39 status report 2009*, CERN-LHCC-2010-004, 2009.

[37] RD39 Collaboration: *Cryogenic Si detectors for ultra radiation hardness in SLHC environment*, Nucl. Instr. and Methods in Physics Research, vol. A 579, p. 775-781, 2007.

[38] V. Eremin et al: *Trapping induced N_{eff} and electrical field transformation at different temperatures in neutron irradiated high resistivity silicon detectors*, Nucl. Instr. and Methods in Physics Research, vol. A 360, p. 458-462, 1995.

[39] V. Eremin, J. Haerkoenen, Z. Li, E. Verbitskaya: *Current injected detectors at super-LHC program*, Nucl. Instr. and Methods, vol. A 583, pp. 91-98, 2007.

[40] E. Verbitskaya et al: *Optimization of electric field distribution by free carrier injection in silicon detectors operated at low temperatures*, IEEE Transactions on Nuclear Science, vol. 49, No. 1, p. 258-263, 2002.

[41] V. Palmieri: *Liquid helium cooled silicon for radiation-hard hybrid superconducting pixel detector*, Nucl. Instr. and Methods, vol. A 426, p. 56-60, 1999.

[42] S. Mueller, "The Beam Condition Monitor 2 and the Radiation Environment of the CMS Detector at the LHC", CMS-TS-2012-042, PhD Thesis 2010.

[43] M. Pomorski: *Electronic Properties of Single Crystal CVD Diamond and its Suitability for Particle Detection in Hadron Physics Experiments*, PhD thesis, Frankfurt, Germany, 2008.

[44] D. Meier: *CVD Diamond Sensors for Particle Detection and Tracking*, PhD thesis, University of Heidelberg, 1999.

[45] A. J. Edwards et al: *Radiation Monitoring With Diamond Sensors in BABAR*, IEEE Transactions on Nuclear Science, Vol. 51, No. 4, p. 1808-1811, 2004.

[46] E. Griesmayer et al: *A Fast CVD Diamond Beam Loss Monitor for LHC*, 10th European Workshop on Beam Diagnostics and Instrumentation for Particle Accelerators, Hamburg, Germany, pp.143, 16 - 18 May 2011.

[47] M. Hempel: *Application of Diamond Based Beam Loss Monitors at LHC*, DESY Zeuthen thesis, 2012.

[48] RD42 collaboration: *Development of Diamond Tracking Detectors for High Luminosity Experiments at the LHC*, CERN Status Report, 2006.

[49] H. Landolt and R. Boernstein: *Semiconductors*, Numerical Data and Functional Relationships in Science and Technology, Volume 17, Springer, 1987.

[50] K. P. O'Donnel and X. Chen: *Temperature dependence of semiconductor band gaps*, Appl. Phys. Lett. 58 (25), 1991.

[51] H. Bichsel: *Straggling in Thin Silicon Detectors*, Rev. of Modern Physics, 60, 3, 663-699, 1988.

[52] C. Canali et al: *Electrical Properties and Performances of Natural Diamond Nuclear Radiation Detectors*, Nucl. Instr. and Meth., 160, 1978.

[53] C. Canali et al: *Electron drift velocity in silicon*, Phys. Rev. B 12, 2265-2284, 1975.

[54] C. Jacoboni et al: *A Review of some Charge Transport Properties of Silicon*, Solid State Electronics Vol. 20, 77-89, 1977.

[55] C. H. H. Wort et al: *Thermal Properties of Bulk Polycrystalline CVD Diamond*, Diamond and Related Materials, 3, 1158-1167, 1994.

[56] V. A. J. Van Lint et al.: *Mechanisms of Radiation Effects in Electronic Materials*, Wiley, New York, 1980.

[57] J. Koike et al: *Displacement threshold energy for type IIa diamond*, Applied Physics Letters, 60:1450-2452, 1992.

[58] M. Aleska and C. W. Fabjan: *Fundamental Physics with Noble Liquid Detectors*, IEEE International Conference on Dielectric Liquids, Coimbra, Portugal, pp.1-12, 2005.

[59] F. Sauli: *Instrumentation in High Energy Physics*, Direction in High Energy Physics - Vol. 9, World Scientific Publishing Co. Pte. Ltd., 1993.

[60] Y. L. Yu et al: *Cryogenic design and operation of liquid helium in an electron bubble chamber towards low energy solar neutrino detectors*, Cryogenics 47, p. 81-88, 2007.

[61] I. A. Kurochkin et al: *Beam loss monitor for superconducting accelerators*, Nucl. Instr. and Methods, Vol. A 329, p. 367-370, 1993.

[62] J. W. Ekin: *Experimental Techniques for Low-Temperature Measurements*, Oxford University Press, 2006.

[63] G. Ventura and L. Risegari: *The Art of Cryogenics*, Elsevier, 2008.

[64] M. J. Buckingham and W. M. Fairbank: *The Nature of the Lambda Transition*, Progress in Low Temperature Physics III, 1961.

[65] Z. Li and H. W. Kraner: *Modelling and simulation of charge collection properties for neutron irradiated silicon detectors*, Nuclear Physics B, Volume 32, p. 398-409, 1993.

[66] P. Norton, T. Braggins and H. Levinstein: *Impurity and Lattice Scattering Parameters as Determined from Hall and Mobility Analysis in n-Type Silicon*, Phys. Rev. BS, p. 5632, 1973.

[67] R. A. Logan and A. J. Peters: *Impurity Effects upon Mobility in Silicon*, J. Appl. Phys. 31, p. 122, 1960.

[68] J. Haerkoenen et al: *The Cryogenic Transient Current Technique (C-TCT) measurement setup of CERN RD39 Collaboration*, Nucl. Instr. and Methods in Physics Research, vol. A 581, p. 347-350, 2007.

[69] http://www.cbwinfo.com/Radiological/radmat/am241.shtml, 07:50, 29.11.2010.

[70] H. Pernegger et al.: *Charge-carrier properties in synthetic single-crystal diamond measured with the transient-current technique*, J. Appl. Phys. 97, 73704-1-9, 2005.

[71] http://physics.nist.gov/PhysRefData/Star/Text/ASTAR.html, 08:00, 29.11.2010.

[72] H. Jansen: *Chemical Vapour Deposition Diamond: Charge Carrier Movement at Low Temperatures and Use in Time-Critical Applications*, PhD thesis, University of Bonn, 2013.

[73] H. Jansen et al: *Temperature dependence of charge carrier mobility in single-crystal chemical vapour deposition diamond*, J. Appl. Phys., volume 113, 17, 2013.

[74] A. Einstein: *Über die von der molekularkinetischen Theorie der Wärme geforderte Bewegung von in ruhenden Flüssigkeiten suspendierten Teilchen*. Annalen der Physik, vol. 322 (8), p. 549-560, 1905.

[75] K. R. Atkins, *Ions in Liquid Helium*, The Physical Review, Vol. 116, No. 6, 1339, 1959.

[76] C. G. Kuper, *Theory of Negative Ions in Liquid Helium*, The Physical Review, Vol. 122, No. 4, 1007, 1961.

[77] A. F. Borghesani, *Ions and Electrons in Liquid Helium*, Oxford Science Publications, 2007.

[78] R. M. Ostermeier and K. W. Schwarz: *Motion of charge carriers in normal He^4*, Phys. Rev. A, 5, 2510-18, 1972.

[79] A. L. Fetter in *The physics of liquid and solid helium*, K. H. Bennemann and J. B. Ketterson (eds.), Wiley, 1974

[80] B. Tabbert et al: *Atoms and ions in superfluid helium*, Zeitschrift f. Physik B 97, p. 425-432, 1997.

[81] B. Tabbert et al: *Investigation of Impurities in Superfluid Helium by Optical Spectroscopy*, Physica B 194-196, p. 731-732, 1994.

[82] T. M. Ito et al: *Effect of an electric field on superfluid helium scintiallation produced by α-particle sources*, arXiv:1110.0570v2 [nucl-ex], 2012.

[83] H. Bauer et al: *Implantaion of Atoms into Liquid Helium for the Purpose of Impurity Spectroscopy*, Physics Letters A, Vol. 137, number 4,5, p. 217-224, 1989.

[84] J. Gerhold: *Helium Breakdown near the Critical State*, IEEE Transactions on Electrical Insulation, Vol. 23, No. 4, p. 765, 1988.

[85] C. Huffer: *Studies of High Voltage Breakdown in Superfluid Helium and SQUID Noise: a R&D Effort to Support the Neutron Electric Dipole Moment Experiment at SNS*, Departement of Physics, Indiana University, 2008.

[86] http://physics.nist.gov/PhysRefData/Star/Text/PSTAR.html, 17:15, 22.03.2013.

[87] S. J. Harris and C. E. Doust: *Energy per Ion Pair Measurements in Pure Helium and Helium Mixtures*, Radiation Research Vol. 66, 11-18, 1976.

[88] W. P. Jesse and J. Sadauskis: *Alpha-Particle Ionization in Mixtures of the Noble Gases*, Phys. Rev. 97, 1668, 1955.

[89] C. Brassard, *Liquid ionisation detectors*, Nuclear Instruments and Methods, Vol. 162, Issues 1-3, p. 29-47, 1979.

[90] R. L. Williams, F. D. Stacey, *Ionisation currents in argon and helium liquids*, Canadian Journal of Physics, 35(8): 928-940, 10.1139/p57-102, 1957.

[91] K. Nakamura et al., Review of Particle Physics, J. Phys. G 37, 075021, 2010.

[92] W. C. van Euten: *Introduction to Random Signal and Noise*, Wiley, 2006.

[93] C. Weiss, E. Griesmayer and C. Kurfuerst: Publication under preparation.

[94] W. de Boer et al.: *Radiation hardness of diamond and silicon sensors compared*, Published in : Phys. Status Solidi: 204 (2007) , pp. 3009, 2007.

[95] N. Wermes et al: *Signal and noise of Diamond Pixel Detectors at High Radiation Fluences*, arXiv:1206.6795v2, 2012.

[96] C. Kurfuerst et al.: *Investigation of the use of Silicon, Diamond and liquid Helium detectors for Beam Loss Measurements at 2K*, Proceedings of IPAC 2012, ISBN 978-3-95450-115-1, p. 1080-1082, New Orleans, USA, May, 2012.

[97] RD42 collaboration: *Proton irradiation of CVD diamond detectors for high-luminosity experiments at the LHC*, Nucl. Instr. and Methods A 426, p. 173-180, 1999.

[98] M. Glaser, L. Durieu, C. Leroy, M. Tavlet, P. Roy and F. Lemeilleur, *New irradiation zones at the CERN-PS*, Nucl. Instr. and Methods A 426, p. 72-77, 1999.

[99] F. Ravotti, M. Glaser and M. Moll, *Dosimetry Assessments in the Irradiation Facilities at the CERN-PS Accelerator*, IEEE Trans. Nucl. Sci., vol. 53, no. 4, pp. 2016-2022, 2006.

[100] https://ps-irrad.web.cern.ch/ps-irrad/BPM40.aspx, 01.12.2012, 11:00.

[101] M. Moll: *Radiation damage in silicon detectors*, PhD thesis, Hamburg, Germany, 1999.

[102] C. Arregui Rementeria: *Design of a Cryogenic Installation for the Test of Beam Loss Monitors (BLM) in Superfluid Helium for the Large Hadron Collider (LHC) at CERN*, Master thesis, Universitat Jaume I, Castellon de la Plana, Spain, 2013.

[103] M. E. Newton et al: *Recombination-enhanced diffusion of self-interstitial atoms and vacancy-interstitial recombination in diamond*, Diamond and Related Materials, Volume 11, Issues 3-6, p. 618-622, 2002.

[104] M. Guthoff: PhD thesis under preparation, 2013.

[105] A. Affolder, P. Allport and G. Casse: *Charge collection efficiencies of planar silicon detectors after reactor neutron and proton doses up to $1.6*10^{16}$ $n_{eq}cm^{-2}$*, Nuclear Instruments and Methods in Physics Research Section A, Volume 612, Issue 3, p. 470-473, 2010.

[106] M. Wolfke and W. H. Keesom: *New Measurements About the Way in which the Dielectric Constant of Liquid Helium Depends on the Temperature*, Commun. Phys. Lab. Univ. Leiden No. 192a, 1928.

[107] S. Ramo: *Currents Induced by Electron Motion*, Proceedings of the I.R.E, Vol. 27, Issue 9, p. 584-585, 1939.

i want morebooks!

Buy your books fast and straightforward online - at one of the world's fastest growing online book stores! Environmentally sound due to Print-on-Demand technologies.

Buy your books online at
www.get-morebooks.com

Kaufen Sie Ihre Bücher schnell und unkompliziert online – auf einer der am schnellsten wachsenden Buchhandelsplattformen weltweit! Dank Print-On-Demand umwelt- und ressourcenschonend produziert.

Bücher schneller online kaufen
www.morebooks.de

OmniScriptum Marketing DEU GmbH
Heinrich-Böcking-Str. 6-8
D - 66121 Saarbrücken
Telefax: +49 681 93 81 567-9

info@omniscriptum.de
www.omniscriptum.de

Printed by Books on Demand GmbH, Norderstedt / Germany